T0252436

How
Climate
Change
Comes
to Matter

EXPERIMENTAL FUTURES
TECHNOLOGICAL LIVES, SCIENTIFIC ARTS,
ANTHROPOLOGICAL VOICES
A series edited by Michael M. J. Fischer and Joseph Dumit

HOW CLIMATE CHANGE COMES TO MATTER

The Communal Life of Facts

Candis Callison

Duke University Press   Durham and London   2014

© 2014 Duke University Press
All rights reserved

Designed by Natalie F. Smith
Typeset in Chaparral Pro by Copperline

Library of Congress Cataloging-in-Publication Data
Callison, Candis, 1971–
How climate change comes to matter : the communal life
of facts / Candis Callison.
pages cm—(Experimental futures)
Includes bibliographical references and index.
ISBN 978-0-8223-5771-1 (cloth : alk. paper)
ISBN 978-0-8223-5787-2 (pbk : alk. paper)
1. Climatic changes—Social aspects—United States.
2. Climatic changes—Political aspects—United States.
3. Climatic changes—Press coverage—United States.
I. Title. II. Series: Experimental futures.
QC903.2.U6C35 2014   363.738'740973—dc23   2014007350

ISBN 978-0-8223-7606-4 (e-book)

Cover: *Walking Dance* (2005) by Siobhan Davies. Photograph
by Gautier Deblonde. *Walking Dance* was created while Siobhan
Davies was on the Cape Farewell expedition to Tempelfjorden,
Spitsbergen, which brought together twenty artists, scientists,
and journalists to experience the Arctic environment.

Printed and bound by CPI Group (UK) Ltd, Croydon, CR0 4YY

For Freya and Kora

Contents

Contents

Acknowledgments

I am deeply grateful to the many individuals who gave of their time and energy in interviews, e-mails, and at the events and conferences I attended as part of research. In bringing together so many groups, my hope is that this book accurately reflects the processes by which climate change has come to matter in such varied contexts and that it creates new intersections for future conversations.

My enduring thanks goes to Mike Fischer and Joe Dumit who provided such amazing mentorship throughout my graduate work at Massachusetts Institute of Technology. The early introduction they both provided to social theory, anthropology, STS, and generative academic conversations offered much needed ballast in my journey from journalism to scholarship. My thanks also goes to Sheila Jasanoff for her generous and close reading of dissertation drafts and ongoing conversations that continue to orient my thinking around STS, science, policy, and publics. Christine Walley also read and encouraged my writing process, providing support on my dissertation committee.

I am greatly appreciative to Ken Wissoker and Elizabeth Ault for shepherding this book through to its final incarnation. As well, I am indebted to two anonymous peer reviewers, one of whom I later learned was Tim Choy. Their deep engagement with this text and insightful comments encouraged me to better articulate just what it was that I thought this amorphous text could contribute to the robust discourse around and about climate change. Any fuzziness or missteps left are of my own making.

A huge thank you to Siobhan Davies for her generosity and permission to use *Walking Dance* (2005) for the cover, a photograph by Gautier Deblonde. *Walking Dance* was created while Davies was on the Cape Farewell expedition to Tempelfjorden, Spitsbergen. During the final drafting of this book, I propped the image up on my desk for inspiration, and it is extraordinary to see it on the cover.

Funding for the extensive itinerant fieldwork that forms the basis for this book came from the National Science Foundation's Dissertation Improvement Grant (#0724753) and from the Center for the Study of Diversity in Science, Technology, and Medicine at MIT. I am thankful to David Jones for supporting this research when he ran the Center.

This book began as a dissertation undertaken at MIT's Doctoral Program in History, Anthropology, and Science, Technology, and Society, but it also has roots in my time in the Comparative Media Studies Program. At MIT, I had the distinct privilege of being taught and supported by incredible faculty: David Kaiser, Stefan Helmreich, Heather Paxson, Hugh Gusterson, Deborah Fitzgerald, Harriet Ritvo, Susan Silbey, Leo Marx, Rosalind Williams, David Mindell, Peter Perdue, Manuel Castells, and earlier in CMS, Henry Jenkins, William Urrichio, Pete Donaldson, Kurt Fendt, and Wyn Kelley. My friends and colleagues from my years at MIT form a community I remain grateful to for past and ongoing conversations: Anne Pollock (who generously read and commented on so much of this book at various stages), Rachel Prentice, Alison MacFarlane, Michele Oshima, Natasha Myers, Zhan Li, Margaret Weigel, Jamie Petriuska, Richa Kumar, Anita Chan, Kaushik Sunder Rajan, Etienne Benson, Esra Ozkan, Aslihan Sanal, Livia Wick, Meg Heisinger, Anya Zilberstein, William J. Turkle, Alexander Brown, Nate Greenslit, Peter Shulman, Timothy Stoneman, Sara Wylie, Laurel Braitman, Nick Buchanan, Orkideh Behrouzan, Xaq Frohlich, Chihyung Jeon, Lisa Messeri, and Rebecca Woods. During the long years in which this book has come about, many other friendships outside of MIT have also informed my research: Michael K. Dorsey, Elizabeth Marino, Tony Penikett, Kirt Ejesiak, Madeleine

Cole, Ed Barker, Cynthia Lydon, Rebecca Zacks, Michelle Stewart, Chris Kortright, Shauna and Michael Bradford-Wilson, and Todd and Maryanne Rutkowski.

In the midst of writing, I found my way back to Vancouver, Canada, and back to my professional roots albeit as a quite different kind of contributor to conversations about journalism. I owe profound and ongoing thanks for the many and multifaceted conversations I've had with so many of my colleagues at the University of British Columbia, especially: Mary Lynn Young and Julie Cruikshank (who both read and commented on early chapter drafts), Alfred Hermida, Kathryn Gretsinger, Kirk LaPointe, Peter Klein, Michelle Stack, Heather Walmsely, Erin Baines, Taylor Owen, David Beers, Pat Moore, and John Robinson. I'm also thankful for the many UBC journalism students whom I have been privileged to work and think with over the past several years.

I am perhaps most indebted to the constant, unwavering support of my husband, who has shared in all the ups and downs of academic life, fieldwork, and book writing with me. My parents, extended family, and the Tahltan Nation have also offered support and encouragement during this academic journey. I'm grateful to be the recipient of so much. Finally, to my kids—thanks for letting Mommy do her work.

Introduction

The existence of the experimental method makes us think
we have the means of solving the problems which trouble us,
though problem and method pass one another by.
Ludwig Wittgenstein, *Philosophical Investigations*

Debates about the nature of climate change often
swing back and forth between what we should do
about it and why the public should (and doesn't) care
(enough) about it. For many, these debates have centered on the stability or certainty of the scientific facts
bound up in the term *climate change.* This book takes
a different approach. It attends to these debates not
only as struggles over complex and evolving "matters
of fact" but also as debates about meaning, ethics, and
morality. Considering climate change as a *form of life,*
this book investigates *vernaculars* through which we
understand and articulate our worlds and the nuanced
and pluralistic understandings of climate change evident in diverse efforts of advocacy and near-advocacy.
Climate change offers an opportunity to look more
deeply at how it is that issues and problems that begin in a scientific context come to matter for wide publics and to rethink emerging multifaceted interactions

among different kinds of knowledge and experience, evolving media landscapes, and claims to authority and expertise.

Climate change poses an inherent double bind for those invested in a variety of stances associated with the communication, journalistic coverage, and public understanding of science.[1] The first half of the bind is this: climate change is ultimately an amalgam of scientific facts based on modeling, projections, and empirical observations of current and historical records found in tree rings, coral reefs, ice cores, sea ice cover, and other forms of data. Acceptance of the premise of climate change requires a fidelity to and trust in scientific methods, as well as institutional processes like the Intergovernmental Panel on Climate Change (IPCC) that collate, elevate, and summarize global research related to climate change. The IPCC is simultaneously a political, social, and scientific enterprise, and what it publishes as the ultimate authority on climate change is based on other institutional prerogatives such as national funding agencies like the National Science Foundation (which also underwrote this research). Not only that, but IPCC reports are also negotiated line by line among country participants and their scientists. With each IPCC assessment report (there are four sets so far, and a fifth is being released as this book goes to press), what can be claimed with certainty about climate change as fact is collaborative, consensus-based, and scientific all at once.

The second half of the bind is that in order to engage diverse publics and discuss ramifications and potential actions, this book argues, based on ethnographic evidence, that climate change must become much more than an IPCC-approved fact *and* maintain fidelity to it at the same time. It must promiscuously inhabit the spaces of ethics, morality, and other community-specific rationales for actions while resting on scientific methodology and institutions that prize objectivity and detachment from politics, religion, and culture. Science and Technology Studies (STS) and anthropology of science scholarship have sought to situate scientific research as occurring within cultures, politics, and institutional frameworks. How evidence is weighted, what expertise matters and when, and what kinds of research get funded all reflect inherently cultural norms and ideals. What this book asks is what role social movements and media play in how facts come to matter for diverse publics and what kinds of debates this opens up about expertise, advocacy, and professional norms and practices in science and media.

In the U.S. context, the question "Why should we care?" is not a straightforward one of translating scientific facts or even of getting all the facts "out there" and into wider public discourse. Nor does science possess the necessary gravitas anymore (if it ever did) to make all or even most Americans care because the facts somehow speak for themselves. If scientific findings were final, few, and powerful enough to demand care and attention from diverse publics, then occasional authoritative media reports might be enough to inform publics and convince them of when (and what) action is required.

Informing and convincing are easily collapsed in expectations of what role media *should* play in societies, and many look to polling data to assess whether or not media are fulfilling such expectations. Though polls vary, Gallup has been asking Americans about climate change since the late 1980s, and its annual reports show public concern ranging between 50 and 72 percent, with a high in 2000 (72 percent) and lows in 1997 (50 percent), and 2004 and 2011 (51 percent). Public polling arguably does little to explain how or why issues become meaningful, and it doesn't show how publics come to care enough to act on an issue. But polling does indicate, particularly over time, how an issue ranks for an aggregate version of diverse publics. Consequently, polls showing how much Americans care about climate change have been and continue to be something of a rallying cry for better and more public engagement, policy changes, and political leadership.

Much has been made about the distinctiveness of American responses to climate change, both in terms of its political response (pulling out of Kyoto, or a refusal to acknowledge scientific consensus in the early years of George W. Bush's presidency, for example), and the public contestation of scientific fact that divides along political and/or religious lines. (In 2008, 49 percent of Republicans were concerned about climate change as compared with 84 percent of Democrats and 75 percent of independents.) It is tempting to write these contestations off as just politics, or worse, as an only-in-America sideshow. Indeed, the most common explanations and reactions include outrage about the perceived lack of public scientific literacy, lack of trust in science, lack of robust media attention to the issue, or some combination of these sentiments.[2] These strong reactions are symptomatic of the double bind inherent to problems of communicating science to wide publics, where facts must be perceived to be produced without investment and yet must be laden with meaning once they leave the scientific context.

Climate change provides exemplary insight into how scientists *and* journalists are negotiating professional detachment and distance, and by extension for publics now forced to sort through claims and counterclaims that take to task scientists and journalists by charging them with bias, exaggeration, or alarmism. Social movements and individual scientists and journalists are attempting to bridge this gap in novel ways, which this book records by analyzing discursive turns that invest climate change with meaning, ethics, and morality. This book analyzes how climate change is being translated into varied vernaculars that make it *a science-based problem with moral and ethical contours*. In this, then, there are analogous applications to conversations about scientific issues and public engagement beyond the American context, particularly in western democracies where publics are expected to become informed and active. Besides western countries, there are also implications here for those who seek to address adaptation and mitigation related to climate change in countries where religious or tribal leaders play key roles in both communicating science to their groups and making their concerns known to wider transnational or international bodies.

Since this research project began, many books have been written about media, science, skeptics, and some of the social movements recorded here (Boykoff 2011; Edwards 2010; Hoffman 2011; Hulme 2009; Mann 2012; Oreskes and Conway 2010; Wilkinson 2012). This book brings these contexts together in order to think with and across societal ideals around science, media, and democracy. However, in contrast to recent attention and energy spent on the psychological processes that might bring about widespread behavioral changes among individuals, this book privileges the role of collectives, shedding light on the structural and societal aspects of how it is that climate change becomes meaningful and what challenges are presented when social affiliations are seen as consequential and constitutive to knowledge. Specifically, it seeks to address how public engagement works if we take seriously the wider commitments of scientists and journalists whose credibility rests on objectivity as a norm and practice (Irwin and Wynne 2004; Jasanoff 2004, 2005, 2010; Schudson 2001, 2002; Singer 2005, 2007; Ward 2004, 2009) and differently constrained social groups who must engage with the conclusions produced by science and articulated through, around, and with media (Choy 2011; Dumit 2004a, 2012; Epstein 1996; Fortun 2001; Fortun and Fortun 2005; Pollock 2012; Rapp 2000; Sanal 2011). In short, it asks who gets to define what a present and future with climate change means, and where, why, and when these differing definitions and epistemologies matter.

What this sets at the center are questions and debates about expertise, advocacy, and adjudicating risk, certainty, and the need for action. Such an approach deeply questions the categories assigned to professions, groups, issues, and concerns, employing an anthropological view suggested by George Marcus and Michael M. J. Fischer (1986) that takes the world as a system, asking how people "constitute their own histories" *and* futures, situate themselves in cultures, and work with and among institutions and their imperatives (Fortun 2003, 180). What Marcus and Fischer and Kim Fortun in her elaboration and experimentation with their work suggest is a conceptual reflexivity that critically engages societal constructs and theories. Such a methodology takes for granted that people must negotiate and act in emergent, complex worlds with an overabundance of information sources, and it employs anthropological tools and analysis in order to draw out differing modes of ethical reasoning in varied communities.

This book uses the multisited ethnographic methods suggested and pioneered by Marcus, Fischer, and Fortun to get inside how climate change becomes meaningful in diverse and specific groups and how this underlying double bind of *maintaining fidelity to science and expanding beyond it* is negotiated by groups that are both central and peripheral to evolving discussions about how to communicate climate change. In so doing, it challenges the ideal of journalists as the public's primary educator, informer, and persuader-in-chief at a time when platforms, norms and practices, and institutions are in flux, and it turns the formulation of public engagement questions around to think more broadly about how facts, meaning, and action are co-produced.[3] Instead of asking why climate change doesn't matter or doesn't matter enough for Americans, or how to improve and foster scientific literacy, this book is based on research that asks how, why, and when climate change *does* come to matter, what that looks like, and what roles there are for journalists, scientists, and social movements among a pantheon of influences and information sources.

The Fieldwork

When I first started working on climate change and public engagement in the early 2000s, I knew only that I was after a general theme that might best be phrased as how science wends its way out to diverse publics, often through media, and who gets blamed when that process breaks down. Roles for journalists, scientists, and publics were very much up in the air, but I knew that I was dealing with ideals that, however flawed, went to the foundation of western democracy and science—that facts and information should and must drive public and political actions (Gans 2003; Schudson 1998). In those early years, what I came across in reading daily news and attending to climate change discussions at MIT were indications of (1) a mainstream discourse in disarray, (2) debates behind the scenes between scientists and journalists about how best to communicate climate change to a disinterested public, and (3) indications of other conversations among indigenous people, religious groups, and business groups going on at the periphery of scientific institutions, policy think tanks, activists, and media.

It's not that alternate discourses were entirely separate, but they certainly weren't conceived as related or constituent to mainstream public discourse either. For example, the 2006 Evangelical Climate Initiative and debates among evangelicals about it made for an interesting couple of articles by the *New York Times* reporter who usually covered religion, and the 2005 human rights case brought forward by the chair of the Inuit Circumpolar Council represented an equally novel and exotic approach to demanding an official American acknowledgment of climate change. Such news articles functioned more like an aside than a serious engagement with climate change science, activism, or policy. Yet the presence of such non-science-focused groups brings alongside the dominant conversations a differently configured and articulated notion of the problem of climate change. For Inuit across the Arctic, it is a direct ongoing and lived experience and one that both threatens cultural practices and brings scientists, activists, industry, and policymakers to their region in anticipation of major environmental change *and* wealth that will likely be generated through resource extraction made possible by a warming Arctic. For American evangelicals, climate change brings into sharp relief both relations with and belief in scientific methods, as well as a call to care for creation (the environment) *and* address poverty and disadvantage likely to increase globally with a changing climate. Climate change then sounds quite different being explained from the pulpit of an American church or from an Inuit elder in a village in Arctic Alaska than it does in the pages of a major American newspaper. This is how I came to devise a research project that looked not only at scientists and journalists but also at social groups that were not expressly or historically related to the issue of climate change.

Using methodologies from anthropology and STS, this book focuses on how five discursive communities are actively enunciating the fact and meaning of climate change:

1. Arctic indigenous representatives associated with the Inuit Circumpolar Council
2. Corporate social responsibility activists associated with Ceres
3. American evangelical Christians associated with Creation Care
4. Science journalists
5. Science and science policy experts

Throughout the 2000s, each of these groups have been heavily engaged with their own group (and other groups to varying degrees) through media, conferences, workshops, events, and personal interactions. Each group is diverse, heterogeneous, and geographically dispersed, requiring multisited fieldwork that focused on collaborative spaces, where articulations, institutional imperatives, vernaculars, and activism, as well as conflict and debate over meaning, effects, ethics, and action are evident and recordable.[4] Although a practice of "itinerant" multisited fieldwork departs from traditional anthropological methodology, it offers some purchase on the way in which facts travel (Dumit 2004a), social movements evolve and form enunciatory communities (Fortun 2001),[5] and media operates as a social practice (Ginsberg, Abu-Lughod, and Larkin 2002).

My fieldwork centered on group and intergroup settings, and I conducted semistructured interviews with the leading voices within each group between 2007 and 2009. Traveling to far-flung parts of the United States from the Arctic to southern Florida, I conducted interviews formally and informally about the issues that members of these groups saw as paramount to the communication of climate change. I was interested in further investigating what seemed to be a mainstream drama full of mishaps, spin, and argumentation, and understanding how scientists and media were working together (or not) to inform the public about climate change. And on the other hand, I specifically sought out communities that dealt with the politics, morality, and ethics related to climate science—those who worked as a minority on the shifting terrain of new and old institutions associated with national and international climate policy and saw the implicit need to operate at both an elite and localized level. My questions asked them to elaborate not just on what they were doing about climate change but also (1) how they perceived public discourse on the issue, (2) the tactical options available through media in flux, (3) the utility of scientific findings and scientific spokespeople, and (4) how they saw themselves situated within and/or outside the climate change conversation.

The five groups were strategically selected both because they engage in the process of enrollment and legitimation (Habermas 1976; Latour 1988, 2005; Weber et al. 2004) and because they represent multiple means of intervention via human rights (ethical imperatives), the church (mobilizing norms and morality), the market (creating incentives and disincentives), mainstream media (mobilization of policy opinion), and science (production of facts and knowledge). With regard to the processes and approaches to educating "the public," they also provide a stark contrast between the dominant dialogue about climate change reported by mainstream media and the often-submerged narratives that are rarely reported on. Together, these groups provide a basis for understanding democratic engagement, conceptions of publics, and the interacting roles of advocacy, science, and media in public discourse.

ICC represents Inuit in Alaska, Russia, Greenland, and Canada who are directly experiencing the early effects of climate change. With the 2004 Arctic Climate Impact Assessment (ACIA), the vulnerability of the northern polar region and the Inuit people became a strongly evidenced part of climate change discourse. The original UNFCCC document drafted in 1992 at the Rio Summit does not list the Arctic as a vulnerable region. ACIA thus represents rather stark evidence that took time to migrate to global policy discussions. What originally got my attention was the 2005 human rights claim that was brought before the Inter-American Commission on Human Rights by Sheila Watt-Cloutier, then international chair of ICC. Though the claim was not submitted formally on behalf of ICC, it was supported by the organization and put the Inuit experience with climate change into wider public discourse in the form of legible rights and claims. ICC's commitment to impart "a human face" to an abstract global problem reframes climate change as a matter of ethics, oral history, decades of experience with its effects, and scientific fact. It also brings to light the long struggle for self-determination across the Arctic—a struggle intimately intertwined with the rush for resources made extractable by a warming climate.

Ceres is a corporate social responsibility group based in Boston that has successfully repositioned and reframed climate change as "climate risk," working to enroll Wall Street firms, insurance companies, and many other national and multinational corporate leaders in their conferences and efforts to push for legislation and regulation related to climate change. While the basic premise and scientific veracity of the research behind climate change is taken for granted, how risk is elaborated, measured, and managed is very much up for grabs. How climatic change will look on the ground for companies with vested stakes and interests in old and new technologies lies at the heart of how they consider climate change–related concerns. Ceres sees itself not as an environmental activist group but as a coalition of investors and environmentally concerned business leaders who seek to implement structural changes in the drivers of American business and investment.

Creation Care is a recent effort to make climate change a Christian concern and responsibility among what's estimated to be 30 million American evangelicals. Traditionally opposed to what's perceived as a left-wing ideology, concern for the environment is being retheorized as a moral and biblical concern, hence the term *creation care* as opposed to *environmentalism*. Science is not the primary basis for their involvement in the issue, though certainly prominent scientists have been involved in their efforts to convince fellow evangelicals. Instead, the role of "messenger" must be performed by those trusted to "bless the facts." Creation Care translates climate change into the vernacular of the group by following biblical and moral dictates surrounding care for the poor and the natural environment.

While journalists and science experts do not conform to the notions of "social group" normally associated with groups like ICC, Ceres, or Creation Care, I am treating both journalists and science experts here as social groups in order to provide rigorous, comparative analysis of how they seek to transform and translate climate change for wide publics and think through issues of action and advocacy. Science experts and science journalists conform to professional norms and practices, and generally they see themselves as part of a larger group of professionals (Gans 1979; Hannerz 2004; Jasanoff 1990; Latour 1988; Merton 1973; Schudson 1978; Ward 2004; Weber, Owen, Strong, and Livingstone 2004). But climate change has compelled many to rethink norms and practices regarding objectivity, detachment, and democratic obligation with regard to engaging the public and persuading them to act on climate change–related facts and information.

In the cacophony of calls for action on climate change, the diverse efforts of these groups to communicate widely has become particularly salient as they are not only competing with other issues but also with conflicting points of view and priorities that have emerged within environmental discourse over the past several decades (Gelobter et al. 2005; Shellenberger and Nordhaus 2005). Each group has varying relationships with the environmental movement, with governments both in the United States and globally, and these intersect in various ways and at differing levels of intensity and collaboration. By studying them together, the language of science, while it may be somewhat diversely articulated in various subfields of scientific research, is shown to be a vernacular with shared idioms, terms, and modes of apprehension for expressing information and views about the natural world (Fischer 2003; Fleck 1979; Irwin and Wynne 2004; Jasanoff 2004; Merton 1973). In the vernacular of all these groups, then, scientific findings are combined with and filtered through other vernaculars, views of the world, everyday life, and democratic citizen responsibilities. While this is most obvious with the three nonscience groups, it is also clear in the ways that journalists and scientists think about articulating science for wider publics and in their discussions of framing and "translating."

The past decade in particular has witnessed the dramas associated with coalescing "scientific consensus" on climate change and attempts to initiate political action (Oreskes 2004a, 2004c). These groups, however, take climate science as scientific and experiential fact and move beyond the debates about the veracity of climate change predictions. The questions they ask are not about the settledness of the science but rather about how climate change might unfold and what ethical and moral responses are required, or even demanded, in response. Most, if not all, must at times contest those who would see climate change as a false or exaggerated claim, but this task remains avowedly at the periphery. Instead, by translating the issue into their own vernacular, these groups and their leading advocates attempt to answer these questions for their own constituents/ audiences *and* wider publics: How do we define climate change for our group? What does it mean? Why does it matter? Beyond the group, the questions begin to shift to who can speak for and about the signs, models, and predictions of climate change, what lingua franca they use, and what constitutes expertise regarding the issue.

Methodology

Changes have occurred in all of these groups since the time when I conducted fieldwork, particularly in the leadership roles, but group sentiments and positions recorded here persist even as the work of the group might have evolved or even changed direction. The contribution of this book, however, is not a report on the positions of these various groups.

Rather it is an approach to thinking about how issues come to have meaning outside the established categories and in/around/through the existing institutions that attempt to manage these issues, both in terms of articulations and actions. Climate change in this analysis is treated simultaneously as object, issue, cause, experience, and body of scientific research, evidence, and predictions—as a linguistic "token" or floating signifier that is given value as an evolving, emergent, overlapping form of life (de Saussure 1986; Fischer 2003; Wittgenstein 2001). It is the way that climate change is articulated, used, circulated, and understood that creates its particular form of life and hence its meaningfulness for individuals and groups.

Ludwig Wittgenstein, in his seminal 1953 work, *Philosophical Investigations*, theorized that meaning is generated socially through use, action, and context and is governed by sets of rules and grammar evidenced through use and action. Wittgenstein's many language games in which he seeks to know, for example, how it is we know we have a brain or that an individual is in fact reading, as Joseph Dumit points out (2003), brings him to culture. We know things because someone we trust has told us or we've read it in a trusted source. This point at which we no longer ask for explanation or more evidence, where "giving grounds, however, justifying the evidence, comes to an end," is the point at which meaning is established—through our *acting* (Wittgenstein [1969] 2008, 204).

Wittgenstein's concept of how meaning is established through action is the underlying methodology that drove my fieldwork and continues to guide my analysis.[6] Such a method takes as its field the collective rather than the individual processes by which concepts come to have meaning.[7] In locating what climate change means (what its form of life is) at various times and places and for various groups, this book throws into question just what climate change is, how "correct judgment" of what the problem of climate change is occurs, and what techniques and moral/ethical codes are used to assess it as a fact requiring action. The meaning of climate change emerges as many assemblages and efforts to explain it compete and collaborate in media and other forums to define the features, rules, and grammars of its evolving and multiple forms of life. Unlike Wittgenstein's games for well-established objects in life like the brain, climate change presents a new conundrum of facts and predictions constantly in motion, requiring translation, clarification, and mobilization as facts requiring actions.

Drawing on Wittgenstein, Michael M. J. Fischer (2003) has argued that techno-scientific problems present as emergent forms of life—replete with ethical dilemmas, the face of "the other," and historical genealogies, "requiring reassessment and excavation of their multiplicity" (58). This explanation of the facets of emergence and the work needed to understand climate change as a problem *in the world* (as opposed to in the lab, or in the policy debate, or in the content of a major news source) has shaped the ways in which my ethnographic fieldwork evolved. I began this research project with a set of questions about whose responsibility it was to get the message of climate change out. But where this book lands is in the midst of ethics, morality, and multiple, sometimes submerged ways of thinking about and being in/with climate change. What bringing together these multiple and diverse discourses leads to are historical and moral contexts for understanding how climate change is understood *and* integrated into ongoing narratives that explain identities, social movement directions and efforts, the production and weighting of knowledge, and structural elements of society that address ethical and societal challenges like fairness, equity, and self-determination.

Climate change then is not a straightforward problem nor is it a standalone fact. It is one in which many sea-level impoverished and wealthy regions of the world may suffer, where polar regions may be transformed, where unpredictable weather may devastate some and elevate others. Thus conceived, *climate change* is a term whose meanings are very much in negotiation among social groupings of many kinds, yet bringing publics into agreement with specific and narrow instantiations of it has been a focus of those involved in policymaking, science, activism, and journalism. Climate change, as this research shows and inherently argues, eludes stability and specificity, both scientifically and socially. We don't know exactly which predictions associated with climate change will come true, nor do we know exactly how to make it matter enough so that everyone begins to act with a future laden with climate change potentialities.

The applied methodology in this case then aims to excavate climate change as a multiply instantiated fact with varying scientific, political, ethical, and moral contours. As an evolving, heterogeneously articulated, emergent form of life, the meanings of climate change are established through attention to multiple discourses, assemblages (institutions, actors, networks), and vernaculars where situated knowledge, advocacy, activism, ethics, and morality become apparent (Dumit 2004a; Fortun 2001; Haraway 1996; Tsing 2005). What this research thus follows are partially submerged narratives and meaning-making processes between and within groups.[8] It tracks the flows of information and sentiments, an evolution of positioning and positions, the emergence of newly configured professional norms, and a struggle for/against the re-inclusion of varied perspectives. This book records parallel processes of meaning-making where the stakes and what success might look like are considered and articulated differently. The ways that climate change is formulated and codified by these groups are constitutive to the understanding, care, investment, and mobilization around the issue.

In some instances, this process conforms to geography and so might be termed "local," but it also conforms to the communal and to the ideological factors that go beyond geography as a unifying factor. In the vein of the Sapir-Whorf thesis on language and worldview, what these groups do is translate climate change into the language of their group: their vernacular.[9] By *vernacular* I mean to signal here the interpretive frameworks by which a term comes to gain meaning within a group and the work of translation that such a term must undergo in order to integrate it into a group's worldview, ideals, goals, perceptions, and motivations to act. The groups described here are variously heterogeneous and fluid, and the notion of a vernacular is meant to describe how climate change has become an issue that a group is concerned about and publicly associated with, as well as how it is discussed, considered, and articulated on its behalf outside of a group. Vernaculars act in myriad ways to provide a kind of boundary-demarcation element for who's in and out of a group and, at the same time, a method for enrollment and membership or identity reinforcement.

In this sense, then, climate change as articulated in graphs, IPCC reports, and peer-reviewed scientific literature is not necessarily universal terminology outside the scientific community. Rather, climate change itself is an object for which the process of how it comes to have meaning is not determined by its scientific origins. I am calling this process "translation," but not in the strictest terms. Wittgenstein's notion of language objects as forms of life provides an alternative framework because, as he points out, even when we speak the same language, grammar, rules, and meaning are still up for negotiation. Meaning cannot be assumed, though it often is; it must be established and agreed upon. I am taking it one step further and arguing that the ways in which meaning is established and evolves also create a set of stakes for the group, group leaders, and if we take climate change seriously, the state of tightly coupled global systems and the nature of risk.

Climate change in its most dire predictions presents a potential reordering and new regime replete with new institutions, ordering of evidence, and political and financial logics (Beck 1992, 2002). The stakes of acknowledging and acting with regard to a future with climate change create community (Fortun 2001), competition (Hoffman 2011), and other responses imbued with moral and ethical imperative (Hulme 2009). That there is a "right" thing to do in response to climate change provokes, for many, an activist stance or lens, and for that reason, this book and the groups it studies form a kind of spectrum of activism and near-activism. Yet it's not just a matter of distinguishing between activist/nonactivist or social group/professional group. Epistemological differences between groups bring into sharp relief how individuals and groups know what they know, which evidence matters and why, what does and should drive action, and what gets categorized as knowledge, politics, morality, or ethics—and why. This inherently striated terrain constrains, creates, and measures potentialities for addressing as well as identifying the sets of concerns associated with climate change. This book then accounts for the ways in which identifying with certain kinds of knowledge production means eschewing morality and politics for some (science, journalism) and requires formulating and/or incorporating that same knowledge differently for others (Inuit traditional knowledge, biblical imperatives to care for creation, risk in financial markets).

For this reason, this book rests on the assumption that addressing climate change requires room for pluralistic conceptions of the problem it poses, replete with reflexivity about where and how knowledge has been and/or is being produced. The implications for communicating science, reporting on science, and moving publics to support science-based policy are indeed enormous. On one hand, pluralism of concept and evidence is what a plethora of new media and information sources facilitate. It is also, in part, what makes journalism a more challenging profession not only because one must navigate competing forms of expertise but because news and feature stories are subject to immense scrutiny, criticism, and counterclaims from concerned audiences with diverse perspectives and expertise of their own. On the other hand, it also means acknowledging and taking seriously what STS scholars have argued are the sociocultural elements of scientific norms and practices and the diversity inherent in the public uptake of scientific facts. Wrestling with these processes of how climate change comes to matter thus reflects both a changed media and information landscape and a changed sense of what constitutes expertise, rationales to act, and problems defined, for many, only *in part* by scientific processes and methods.

While I don't deal explicitly here with skepticism or the production of doubt (Hoggan and Littlemore 2009; Lahsen 2005b; Oreskes and Conway 2010), its specter hovers variously as a reference point, rationale for action, and factor of political partisanship. Each group must contend with the persistent existence and circulation of skepticism regarding climate change predictions. In contrast then to Naomi Oreskes and Eric Conway's "production of doubt," this book focuses on the "production of care" (Dumit 2004a; Fortun and Fortun 2005). It records how groups negotiate with the central problem of how to frame a long-term, uncertain issue with a wide spectrum of possible outcomes so that immediate action is required. Traversing the margins and/or teetering on the precipices of alarmism and lack of perceived objectivity are primary challenges for journalists and science experts. While ICC, Ceres, and Creation Care, by comparison, are much more comfortable with moral and ethical claims given the explicit codes that guide and differentiate their groups, their primary challenges lie in the processes of translation and mobilization. In other words, the fact of climate change does not advocate, on its own, for taking action now or in the future—it is the presentation of the facts, their socialization and communality, and inherent moral and ethical dictates that determine the need for immediate action.

This book thus argues that alongside the dominant discourses through media in the midst of transformation, social networks and affiliations provide a vital translation of science in varied vernaculars such that climate change becomes invested with meaning, ethics, and/or morality. This translation, however, is never without friction, nor is it homogeneous or monolithic (Benjamin 1968). Rather, the process of translation into vernaculars enrolls an assemblage of institutions, material training, disciplining mechanisms, and modes of speech in order that articulations might emerge regarding what climate change means for diverse publics and social groups (Fischer 2003; Fortun 2001; Jasanoff 2005; Tsing 2005; Wittgenstein 2001).

Context

The period in which I undertook the intensive fieldwork for this book is distinct because of where it sits on a number of different timelines. For example, if one looks only at American policy developments, or only at

IPCC reports, or only at public polling data and/or media reports, 2007/2008 represents a culmination of efforts begun much earlier. It is also a midpoint between when climate change discourse was marked by frustration with lack of public and media engagement, and the current moment marked by both fatigue and newly emergent forms of activism. Where the early 2000s looked to engage the public through more and better reporting in the mainstream media, the early part of the 2010s has seen a decline in reporting by 40 percent since 2009 (with a brief exception related to climategate, which I will explain later), the failure of multiple UNFCCC Conferences of the Parties to reach consensus on global policies, polling data where increasing numbers of the public who see climate change reporting as exaggerated, a proliferation of blogs and other online information sources, and new forms of media activism by groups like 350.org (led by Bill McKibben).[10] The year 2013 marks a new turn in American policy with President Obama's new national climate action plan. When he launched it, he asked Americans to "speak up for the facts" in their social groups and communities, obliquely referencing the still lingering skepticism and resistance to climate policy in the U.S.

Situated in between these two periods, my fieldwork occurred during a time in which the U.S. media reported on climate change more than in any era previously or in the years since. Media coverage peaked in 2007 in part due to the release of the fourth IPCC assessment reports, which included a major public relations and media initiative, as well as unequivocal statements about the role of humans in climate change. But 2007 also reflects the momentum of years previous in which, as one conference speaker I heard explain: "Hurricane Katrina blew the door down, and Al Gore walked through it." Hurricanes Katrina and Rita in 2005 caused immense destruction and loss of life in the Gulf of Mexico and were strongly linked to climate change due to a study published by MIT professor Kerry Emanuel in Nature on August 4. In 2006, Al Gore's phenomenally successful documentary, An Inconvenient Truth, was released and went on to win an Oscar. Gore was also awarded a Nobel Peace Prize along with the IPCC authors in 2007 for their work in putting climate change on the public agenda.

According to Gallup, which has the longest running polls on global warming, American public worry grew slowly during the period of high media attention from 2006 to 2008, but it did not exceed other peaks in 1999/2000. Beginning in 2009, public concern shrank back to the lows of 2004, only to begin a slow rebound in 2011. What's remarkable in this yo-yo, however, is that public worry over climate change began at 63 percent in 1989 and has stayed between 50 and 66 percent for most of the intervening years. Other polls have found much higher and lower public concern about climate change, but Gallup remains something of a baseline because of the longevity, consistency, and regularity of its polling practices on this issue.

In response to the range of polling results, the dominant dialogue about climate change communication that I heard consistently leading up to and during my fieldwork revolved around how to get science across to the public, how to frame it, how to "sell it" to the media, and how to get editors to publish more stories until the American public can't help but care, pay attention, and regurgitate the facts of anthropogenic climate change on a global scale. Often, when I introduced my research as concerning "the communication of climate change to Americans," I was answered by some kind of exuberant comment like "We need that!" This usually would be quickly followed by questions about whether my findings would help solve what seems like an insurmountable problem, namely, educating or informing the public.

Diagnoses of the problems plaguing public engagement with climate change abound.[11] Many blame the presence of industry-funded skeptics who continue to tout climate change as theory (instead of fact) replete with uncertainty, contrasting it with an ideal of "settled science" that would warrant action. The response from most in science, science policy, science journalism, and environmental advocacy has been to either reaffirm the veracity of climate change–related facts or increase the amount and decibel level of activism in order to combat this "production of doubt" (Oreskes and Conway 2010). Skeptics have countered with accusations of unnecessary alarmism and a parade of experts that defy or ignore the core of peer-reviewed factual claims that affirm the basic tenets of climate change. In 2009, just after I wound down my active fieldwork, skeptics nearly gained something like an upper hand as a result of what's been called "climategate," an incident where e-mail accounts of international climate scientists associated with the University of East Anglia in the UK were hacked and their e-mails made public.

Many feared that climategate had dealt a serious blow to efforts at eradicating skepticism and moving toward personal and political change in the United States. The e-mails revealed that scientists wanted to make sure climate change was not questioned and seemed to imply a bias in the peer review process. But multiple formal inquiries into climategate cleared scientists of wrongdoing and exonerated the peer review process, finding that although the scientists had perhaps talked disparagingly of others and sometimes sounded like advocates for climate change, they did not exclude or enact bias based on their views that climate change must be addressed. Even with climategate now receding in memory, public engagement with climate change remains inconsistent and on the lower end of the spectrum of American concerns, or worse, if one takes into account the rising numbers of those who think climate change findings have been exaggerated. Consequently, what remains a favored perennial target for advocates, scientists, *and* skeptics are "the media."

In a 2011 *Rolling Stone* essay, Al Gore provides something of a laundry list of these persistent critiques of the role of media related to climate change. He lambasts media for not revealing skeptics as charlatans, not representing scientists as having achieved consensus, and not doing more to encourage the public to care and agitate for political action and policy changes. As science journalist Keith Kloor summed it up in a blog post shortly before Gore's article came out: "If there is one deeply held sentiment in the climate debate that is shared by bloggers and commentators of all stripes, it is this: journalists suck." Conversely, at a panel hosted by the UBC Graduate School of Journalism (where I am a faculty member) during the American Association for the Advancement of Science annual meetings in Vancouver in February 2012, one journalist described writing about climate change as akin to "parking your car under a bunch of starlings," referring to the robust online blogging and commentary communities that are likely to respond to any story journalists write about climate change. In this continually refiguring relationship between journalists and their audiences, reporting on climate change has put entirely new demands and stressors on a profession devoted to reporting "just the facts." Interpretation, representation, and adjudication of risk and expertise are all open to public scrutiny and criticism.

When I began my research, there were public laments from climate scientists for a Walter Cronkite figure who would have been able to alert the general public to the urgency of the issue. Cronkite was a longtime CBS television and radio anchor (1962–81) who reported on major events like the assassination of President Kennedy, the Vietnam War, and the landing of the Space Shuttle. He is seen as a man who earned the trust of the American public, and were he still on air, he would be able to lend credibility and concern to climate change. This is a view that doesn't take into account changes in media, publics, and science, where a plethora of sources for information and commentary often collide with a scrutiny of independence and demands for transparency. For example, in 2012, several journalists were targeted by at least one think tank through Freedom of Information (FOI) requests for e-mail exchanges with scientists at public universities. The claim based on subsequently released e-mail exchanges was that journalists were biased in their support of climate science and unable to objectively adjudicate or represent the claims of climate skeptics.

Embedded within these passionate indictments against journalism or "the media" is a key democratic ideal: the expectation that the average citizen will access and, if need be, pursue the required information in order to make rational decisions about issues of concern (Habermas 1962; Schudson 1998; Terdiman 1990; Warner 1990). Media offers a key conduit for awareness of new information, but it is also subject to a range of factors related to professional norms and practices such as ranking, what's deemed new and news, and other production exigencies. With the rise of new media, the hope has been that with new sources for information and more direct access as well as participation by users (Jenkins, Ford, and Green 2012; Rosen 1999; Singer et al. 2011), more information might lead to more and better democratic involvement. In the case of climate change, many have pointed instead to the confusion that new information sources seem to have generated.

Misinformation, offered either by underinformed journalists or those who intentionally twist or obscure facts to undermine or deny widespread scientific consensus on climate change, puts a serious "wrench" in the idealized process of facts leading to public response and/or action. Certainly, political calculations have always involved the proverbial "interest groups" and accounted for "spin," but science has usually been portrayed as above the fray. The ideal operating in the political, media, activist, and policy spheres is that science should be impartial, nonpartisan, and objective, providing data and facts regardless of creed or association (Bush 1945; Merton 1973).[12] Journalism, too, in its ideal American form, is a purveyor of fair, balanced representations of facts that drive action and responses within public and democratic institutions.

Democratic and scientific ideals share this in common: that the discovery of objective facts and the dissemination of that information will drive action. But the line between information and action is anything but straightforward, and more often than not, information must traverse not only the vagaries of media channels for mass communication but a diversity of meaning-making, ethics, and morality. In an era of immense upheaval in the structures of media that have come to define democracy, the process by which information gains wide public attention has only become more complex and difficult to ascertain. What democratic, journalistic, and scientific ideals often leave out are the ways in which facts are produced and circulated (Jasanoff 1990, 1991, 2004, 2005; Latour 1988; Latour and Woolgar 1979; Traweek 1988) and the vast social networks and affiliations where individuals negotiate and determine positions, identity, and meaning, often in conjunction with a wider group process (Dumit 2004a; Fischer 2003, 2009; Fortun 2001).

This book argues that more information is not what is required to fully engage the public on the issue of climate change.[13] Instead, the applications of meaning, ethics, and morality—linking to what people already care about through a process of articulation and translation, plays a central role in public engagement with complex scientific issues (Jasanoff 2010). Such a formulation challenges models of the public understanding of science which are predicated not on what science means for diverse publics but rather on scientific literacy, efficacy of communication, trust in science, and achieving public comfort with uncertainty. In short, the conclusions based on this research suggest these models need to be rethought so that information is not divested of the process of socialization and meaning-making inherent to the public adoption of facts as matters of concern (Jasanoff 2004, 2005; Latour 2004a). Specifically, it tackles underlying issues that are constituent to considerations of public engagement with climate change and about which much has been written: (1) public understanding of science, (2) media changes, and (3) the use of expertise by media.

Publics, Understandings, and Science

Models for improving the "public understanding of science" usually assume several basic elements: an authoritative stance for science, a set of facts that can be communicated, the need for a democratic public to know, and a measurable lack of scientific literacy among the general public. True to the ideals discussed above, public understanding models reflect the sentiment that if the public only knew more facts, or "all" the information, they would be compelled to act on the ramifications and potential impacts, exercising their duties and obligations as citizens to undertake collective action and activism through political and practical means. Irwin and Wynne (2004) point out that such models invariably put a homogeneous body of science up against an only slightly less monolithic body of nonscientific understandings, denying each the "wider commitments and assumptions" that a social *and* technical framework for analysis might provide.

Sheila Jasanoff builds on this observation in order to specifically challenge notions of scientific literacy: "The greatest weakness of the 'public understanding of science' model is that it forces us to analyze knowledgeable publics in relation to their uptake of science and technology rather than science and technology in relation to their embeddedness in culture and society" (2005, 271). Thus the nonscientific public is always at a loss in terms of their knowledge of how or why a particular issue may or may not be a matter for wide concern, nor are issues linked to existing concerns, beliefs, or other kinds of knowledge. This presumption undergirds the call for better science literacy, but it also refuses an understanding of science as an iterative process and as a nonmonolithic mode of apprehending the natural world with varied approaches and sometimes competing subfields. For example, one journalist described the latter issue as a key problem in the early years of reporting on climate change, where scientists working on the ozone hole were likely to disparage their colleagues down the hall working on the still newly emerging problem of climate change. This anecdote captures something about the institutional apparatuses that govern scientific research, where problems take time to become widely recognized through peer review processes that oversee research funding and publication of results.

In her analysis of several historical moments when scientists have disagreed, Oreskes (2004b) has pointed out that differing scientific fields see problems through methodological lenses that inform how they define the problem and address it. At the same time, scientists are enmeshed within particular sociocultural institutions with their own priorities, parameters, and politics. But it's not just the old adage that "to a carpenter with a hammer, every problem looks like a nail," meaning that we use the tools and approaches we have to solve the problems that are apparent to us. Rather, the proverbial "nail" must be articulated within existing frameworks that prioritize some evidence over others and reveal deeper cultural ways of knowing and seeing the world. Epistemology in science can be seen in varying shades then as well (Knorr-Cetina 1999; Nowotny, Scott, and Gibbons 2001).

In terms of the public, Jasanoff proposes "civic epistemology" as a conceptual tool to acknowledge the multiple understandings that publics use to both adjudicate scientific evidence and understand it within the frames of history, culture, prior interpretations, and alternate forms of establishing meaning. Simply put, evidence is weighted differently in different cultural contexts. In Jasanoff's study, she compares Germany, the United Kingdom, and the United States where starkly different national contexts drive very different responses to the use and adaptation of biotechnologies for consumer use, including the hot-button issue of genetically modified foods. Jasanoff arrives at the concept of civic epistemologies as a way of accounting for tacit collective knowledge—ways by which publics evaluate scientific claims. Rather than assuming authority for science when facts are so labeled, civic epistemologies help to assess the means by which knowledge comes to be perceived as reliable and authoritative.

This book thus investigates and theorizes the paths and subpaths by which civic epistemologies in the U.S. are evolving and forming around the problem of climate change. The mechanisms that support the evaluation of claims are not merely based on their status as accepted scientific fact. In some cases, who is making a claim matters almost as much as what the claim is. For example, "the messenger," as Creation Care representatives term it, is highly influential in their groups' adjudication of the veracity of climate change. Relatedly, what a claim sounds like in the vernacular of the group takes precedence in articulating a rationale to act. Ceres's use of the term *climate risk* mobilizes a response to an environmental problem that rests more on fiduciary obligations than on care for the natural world. Labeling something a "risk" in a business vernacular means it must be managed and accounted for in order to satisfy investors and insurance companies as well as business imperatives and ideals. Vernaculars elicit trust as well as action from a group's members. Studying a group like Ceres or Creation Care makes this plainly obvious, but what Jasanoff's work also points to is that we may not recognize the underlying cultural assumptions in our contexts except in comparisons to others.

Critiques of public understanding models and the notion of civic epistemology primarily focus on scientists' and policymakers' efforts to educate or engage the public and regard media as one of several mediating factors. This research builds on these interventions and concepts, but also considers mainstream media a key mediator in dominant conversations about climate science. In this framework, science and media are considered two idealized forces constituting the overarching democratic ideal in America, and both pose significant problems in terms of how to consider "the public." Information in both ideals is the key interface by which understanding and democratic action related to science should proceed. Information about climate change has tended to be scientific, although economic predictions and models have also begun to compete for public attention as an application or vision of what aggregated scientific findings portend.

Climate change as a problem originating in scientific and science policy contexts conforms to a mode of knowledge and fact production that divests itself to a great extent of an ethics or morality applicable to everyday living. It is this "native" version of climate change that environmentalists and science journalists (and notably Al Gore) have drawn upon through the use of experts, graphs, and other kinds of evidence. Science, providing an objective detached set of facts as evidenced through peer-reviewed research, is the substance of their narratives and efforts to persuade. The implicit argument is that these are, after all, a set of facts that demand action and a transformation of society. This, in a sense, forces into the foreground the relationship of their audiences with science in general, raising questions about its trustworthiness, particular epistemic variety (for example, paleoclimatology versus atmospheric modeling), and scientific literacy. It takes for granted a scientific mode of apprehending the natural world and glosses over the nature of science as iterative process in favor of textbook facts.

What gets passed over in this narrow sense of climate change is that these conclusions and predictions, while scientific in origin, have the potential to thrust much larger questions into the foreground such as the relationship of the individual to community and to nature. In a mainstream culture dominated by consumerism, celebrities, and market research, climate change cuts to the core of who and what human concerns are and how they are mediated and moralized. It enables questions beyond what the realm of science offers: What is our relation to each other, locally and globally? What is our relationship with the earth—an entity or bounty that we have taken for granted through much of the industrial age? What does the future look like if our impulses and choices remain unchecked? These questions are what energize some of the social groups I studied and confound some of the professionals I interviewed.

Climate change challenges people to see themselves as part of global environmental, industrial, and capital systems, and in many ways it demands a co-articulation of how to locate oneself in a larger collective. Literacy[14] regarding climate change might thus be understood as an aspect of meaning-making—as finding one's feet amid ambiguities, navigating new categories of knowledge and expertise, and wrestling with the notion that impacts and actions can be measured, assessed, and assigned. What this research brings to the fore then is the way publics engage in a version of what Joseph Dumit (2004a) calls objective self-fashioning, but at the level of group and group identity. Group leaders and members negotiate with the scientific facts, setting them at the intersection of their own belief systems and ethical imperatives as well as other cultural inputs outside their group. For the groups working outside the mainstream discourse, this process is very clear. They bring climate change into the realms of both fact and concern, investing meaning in information and facts so that the need for action is not a next step but a constitutive part of moral and ethical codes. So, for the Inuit, climate change is infused with traditional knowledge, and experiencing the effects directly becomes the standard-bearer of evidence and the driver for communicating globally. For Ceres, economic growth and sustainability are in jeopardy as long as the risks associated with climate change are not addressed, and acknowledging this opens up a host of opportunities for new markets and commodities. For evangelicals, it is both a new understanding of their Creator and a call to consider those least fortunate that are the moral underpinning for addressing climate change. Science in all of these cases is a partner and not the sole evidence used to persuade first the group, and later a more generalized notion of a public, of the need to act.

For scientists and journalists, this process of negotiating an ethical stance with regard to climate change is a much differently fraught process. Journalists, like scientists, are enmeshed in cultures of professionalization, institutions, structural and hierarchical relationships, contexts, histories, technologies, and funding/commensuration issues (Benson and Neveu 2005; Blum, Knudson, and Henig 2006; Boczkowski 2004; Cook 1998; Fiske 1996; Gans 1979; Gitlin 1980, 2002; Hannerz 2004; Herman and Chomsky 1988; Kovach and Rosenstiel 2007; McChesney 1999; Schudson 1978, 1995; Winston 1998). Yet facts and predictions regarding climate change as well as the production of doubt have compelled many to find ways to speak to and/or for the need to act. In this book, I use the term *near-advocates* to refer to those like scientists and journalists who are bound by professional norms of objectivity, independence, and distance, and yet choose to articulate (in varying ways and to varying extents) the need to address the wide range of predictions and risks associated with climate change.

In bringing together these multiple discursive practices and processes of articulation, what slowly becomes evident are the epistemological differences: the ways in which climate change and its subjects are constituted, what counts or is debated as rational or scientific, and the ways in which relations of democracy and knowledge when it comes to technology and science are "always up for grabs" (Dumit 2004a; Ezrahi 1990; Foucault 2003; Haraway 1991, 1996; Jasanoff 2005; Jasanoff and Martello 2004). I am using the term *vernacular* in this book not only to differentiate the ways in which climate change is multiply instantiated as heterogeneous, interconnected, and related forms of life, but also to point to the ways in which discourse is a material-semiotic practice experienced and generated through multiple mediated and non-mediated means and human and nonhuman participants.

Experts and Media

This book seeks to open up assumptions, expectations, roles, and processes related to the interplay between media, democracy, and science. In my analysis, all of the groups—not just journalists and scientists—play various and often key roles in the translation and transformation of scientific issues for diverse American publics. Yet media and, more specifically, journalists are perceived and idealized as a "fourth estate," a concept built on the ideal of widely available, independent, objective reporting that holds the state and all its systems accountable. In the last two decades, this ideal has been assailed by the decline and fragmentation of traditional audiences, changes within journalism, restructuring and mergers of major media companies, and the rise of new forms of media—all are acting to shift and transform the role of media in American democracy. All forms of media are in a prolonged state of flux as cable news, the opportunities and demands of 24/7 reporting, the shift to entertainment-oriented news, and the rise of the Internet play contributing roles in the difficulty of addressing complex issues through current news formats. The depth of media influence then is much more difficult to gauge, as is the notion of the general public they once addressed (Castells 2009; Jenkins 2006a, 2006b; Singer et al. 2011).

In addition to this complex and unpredictable pattern of change, part of the gauntlet journalists face in the task of "educating" and/or "informing" the public about science is structured by long traditions within the industry itself. Science is primarily reported at the national level. Journalists and editors thus face the issue of relevance at the local and regional levels, as well as the not insignificant problems of how to explain scientific concepts and adjudicate scientific language and expertise for general audiences. As this book narrates, journalists and the scientific experts they draw on struggle with how to negotiate the imperatives that arise from scientific findings and projects, and further, how to make it relevant to their varied and broad audiences. In addition, they are dealing with structural problems confronting the now multiple platforms of media—those of eroding audiences, ownership consolidation, and declining profits that force the end of special science sections and science reporters. But the daunting task of engaging heterogeneous publics gets at something much deeper, which I have earlier called the democratic ideal.

Several scholars have tackled the relationship between information, the printed word, and democracy (Habermas 1962; Terdiman 1990; Warner 1990). Michael Schudson, a historian of journalism, describes the relation between information, citizenship, and media as an only recently arrived cornerstone of American democracy (Jenkins and Thorburn 2004; Schudson 1998). Early Americans who partook in civil society were white men of a certain social standing who raised their hands to cast their votes and thereby affirmed the social rankings of their time. This civility was followed by an era of raucous party politics that saw voters turn out in high numbers because they were often paid for their vote by their party of choice, and the spectacle they would encounter on voting day affirmed a sense of community loyalty to party, and of course it was thoroughly entertaining. Voting reforms during the Progressive Era (1890s–1920s) introduced the vision of a free-thinking, literate individual voter who subverted the power of the political party, weighed the merits of candidates, and cast a secret ballot—an ideal Schudson terms the "informed citizen." This ideal persists, but media and hence access to information are in the midst of rapid transformation, and what kind of landscape this transformation will produce in terms of democracy is still anyone's guess (Boczkowski 2004; Boler 2008; Jenkins 2006a; Jenkins and Thorburn 2004; Singer et al. 2011).

Since the rise of the Internet in the late 1990s, media has been in an increasing state of flux. Some see an ultimate destabilization of broadcast hegemony through the fragmentation of audiences and a greater number of sources that includes citizen journalism or nonjournalists' reporting (Anderson 1991; Benkler 2006; Boler 2008; Castells 2000, 2003, 2005; Gitlin 1998; Habermas 1962; Jenkins 2006a; Schudson 1995; Singer et al. 2011).[15] Recent reports indicate a more complex patchwork where advertising and news content are being decoupled as radio, television, and newspaper audiences steadily decline, but traditional newsrooms, particularly newspapers, still account for most news content online.[16]

As a result of these changes, Schudson began observing that the informational citizen was beginning to give way to a "monitorial citizen" who, overwhelmed by the onslaught of data, due partly to the rise of new media, engages in surveillance more than actual information gathering. Schudson compares this form of citizenship to parents at a pool who keep an eye on everything, ready to jump into action should the need arise. Theories of collective intelligence are in part based on what Schudson calls the monitorial citizen, as are newsroom fears about audience decline—and the rise of new media technologies play a leading role in the new form of citizenship Schudson describes (Benkler 2006; Jenkins 2006a, 2006b; Jenkins and Thorburn 2004; Levy 1997). Henry Jenkins, building on notions of collective intelligence, has recently argued in his book with Sam Ford and Joshua Green (2012) that news and other content derives its value as a cultural resource within networked communities in part because users share it with one another. This adds more texture and higher stakes to the notion of monitorial citizenship and imparts a quite differently configured challenge for journalists.[17]

What new media technologies and the climate change debate put front and center for consideration is the dissemination of expertise, the tone/tenor of public debate, and the role we expect media to play in adjudicating both expertise and debates about climate findings. Debates about the role of expertise and journalism in American democracy began in the 1920s, long before the rise of blogging and social media. As media undergoes massive changes, these debates have been revived in order to undergird notions of public and participatory journalism (Gans 2003; Glasser and Craft 1998; Merritt 1995; Munson and Warren 1997; Rosen 1999). Walter Lippmann, writing in 1922, saw journalists as the link between powerful insiders and the general public, whom he saw as largely ignorant, self-absorbed, and incompetent—and in need of experts able to make decisions on their behalf in an increasingly complex society.[18] His later work only cemented this pessimistic view of both the public and democracy, prompting a response from John Dewey in *The Public and Its Problems* (1927). While admitting that Lippmann's indictment was almost entirely correct, Dewey saw the public as ultimately capable of a negotiation with facts, rational thinking, and action.

Vacillations between Dewey's and Lippman's stances hover over any analysis of climate change communication. Climate science has continually run aground when experts have been pitted against other experts—in all forms of media, and most particularly on blogs and websites. With an evidenced declaration of scientific consensus, the vast majority of climate-related scientists seemingly prevailed (Hoggan and Littlemore 2009; Oreskes 2004a, 2004b, 2004c; Oreskes and Conway 2010). Skeptics, however, still continue to counter with petitions, conferences, and other media-oriented interventions. Expertise then will likely continue to play a leading role in unraveling the climate issue. Making, critiquing, and collaborating for scientific, economic, and policy projections requires a mix of in-depth expertise—even scientists and economists have been challenged as to how to undertake the kinds of interdisciplinary collaborations that climate policy demands. How then to puzzle through the problem of expertise in an age of proliferating media technologies and sources for information, commentary, and analysis?

STS scholars have repeatedly shown the ways in which expertise is constituted and deployed from and within specific contexts that are inherently social, institutional, political, and historical (Collins and Evans 2002; Gieryn 1998; Hilgartner 2000; Jasanoff 1990, 1991, 2003; Lahsen 2005b; Lynch 1998; Miller and Edwards 2001; Nowotny, Scott, and Gibbons 2001; Oreskes 2004b; Shapin 1995; Walley 2004). Asking who counts as an expert and what knowledge is included or excluded are two key starting points from which to proceed through the maze of expertise that dominates science and policy arenas. Jasanoff (2003) has characterized the United States as a litigious society with a contentious environment for the content, discourse, and evaluation of expertise. This analysis helps in part to explain the debates over climate change in the United States, which are and continue to be strikingly different than in most other developed countries. This difference, however, is not just bound up in the contentious nature of public debate. It is also the American desire for objective, scientific facts—objective and scientific often being conflated as the same thing in common parlance. Yet as Irwin and Wynne write, it is an impossible ideal to achieve: "The 'facts' cannot stand apart from wider social, economic, and moral questions even if rhetorically they are often put forward as if this were the case" (2004, 3). Jasanoff argues further that the objective or "view from nowhere" stance is itself culturally specific.

The notion that facts can and do stand on their own, apart from where, how, and by whom they are produced, is fundamentally embedded in the idea of providing enough information and verification of facts for an informed citizen. But a focus on "just the facts" doesn't entirely blot out attendant social, economic, ethical, or moral factors. It does not account for the interacting, overlapping processes of media, science, and policy, nor does it consider the commitments, knowledge, and voicing of a heterogeneous public (Haraway 1989, 1991; Jasanoff 2004; Latour 1991; Rabinow 1992; Shapin and Schaffer 1989).

Many STS authors have pointed out that as Oreskes puts it, "scientific proof is rarely what is at stake in a contested environmental or health issue" (2004b). Instead, science becomes a factor and/or catalyst in often boisterous debates revolving around political, moral, and ethical claims. Dealing with these claims lies outside the realm of both science and science journalism's traditional focus on gathering and communicating facts. Ethical debates inherently involve ideas about democracy, community, and social networks that cannot be written off as either ideological or an irrelevant sideshow to confronting how it is that facts come to matter. It is in this space that near-advocacy must be negotiated by scientists and journalists.

Including ICC, Ceres, and Creation Care in this research offers a clear departure from the mainstream narrative of the failure of media to report adequately on climate change, arguments over the veracity of climate change as scientific fact, or various policy options that should be explored. Rather, by incorporating climate change as an undisputed fact, these groups embark on something like *a relationship-building and translation exercise with the scientific facts*. Meaning, ethics, and morality are negotiated within the group's vernacular, history, identity, and epistemology. In so doing, they break through, go around, and/or transform the expertise-laden broadcast modes of transmitting information to the public. Information in this sphere is never without a position, narrative, perspective, or spokesperson. The form their strategies and efforts take largely depends on how convinced their own group is of the importance of the issue, and it's from their group's perspective that they hope to influence or contribute to a much larger shift among a generalized notion of "the American public."

How much then does the issue (or the science) change once it moves onto the terrain the group inhabits? Or rather, *how much do epistemological differences matter*? This is no small point when considering any of the groups this book analyzes and the kinds of vernaculars they mobilize on behalf of climate change. Social movement theory and ethnographies of social movements have generally focused on how groups coalesce around issues, rather than the reinvention of groups to address new crises that arise independently of the issues or category of issues the group originally formed to address (Epstein 1996; Fortun 2001; Melucci 1989, 1996). Observations about how groups form still remain helpful for this analysis. In particular, Pierre Bourdieu (1991) theorized that a group comes into being with the election of and personification through a spokesperson, but in this case, spokespeople emerge not in order to bring the group into being but to bring the issue into being for the group by investing it with meaning. The group becomes a "spokesgroup" for the issue. The relevant questions then morph from what it is to who can speak to/for/about climate change? Does it matter from which perspective: human, economic, ethical, moral, scientific? And what constitutes expertise regarding climate change?

What becomes evident in posing these questions is that notions of expertise become morphed as the definition of climate change begins to emerge more fully as a multiply instantiated form of life. This book shows the ways in which material assemblages—institutions, conferences, training opportunities, websites, press releases, and media coverage reflect struggles over how to define the problem of climate change and thus become an expert—particularly in reference to its impact, ramification, and possible solutions. Vernaculars and articulations form the infrastructure for climate change as an experiential, pluralized form of life. And as media has expanded through blogging and other forms of 24/7 reporting, the evolution of these articulations and morphing of expertise becomes that much easier to access and track—whether publics inside and outside these groups actually will or do follow suit as invested, engaged "monitorial citizens" is a problem to be examined in the future and one hopefully informed by the research, analyses, and methodology put forward here.

Experimental Contributions

This book brings together and contributes to scholarly literature in media studies, STS, and anthropology. It rests on an interdisciplinary perspective that seeks to trace the contours of public engagement with climate science in an American context, and it provides a deep sense of the particularities of how climate change comes to matter in diverse groups operating within American and global science, media, and politics. Studying groups together and simultaneously provides a way to understand and critique what modes, practices, and systems there are for public, democratic discourse regarding complex scientific issues. It allows for a multifaceted investigation into how expertise is crafted and what role there is for meaning-making, morality, and ethics in relation to science. Given the structural and technological changes afoot for media, dilemmas presented here in relation to journalism and science expertise are quite likely to grow and diversify, and groups like ICC, Ceres, and Creation Care can be expected to become more significant rather than less in efforts at public engagement.

Each of the groups studied together in this book could be a research project on their own. One of the challenges then is to maintain fidelity to the research I have conducted with each of the groups. Bringing them together generates perhaps a more shallow view than one would have were they studied solely and for a longer period of time per traditional ethnographic fieldwork practices. I prefer, however, to think of this as a kind of "jeweler's eye view," where one is able to tack back and forth between the micro and macro views of the object of research and study (Marcus and Fischer 1999). The synthesis is essential for macro views and attends to the dictate of following systems, institutions, processes, and interactions such that assemblages, modes of speech, and material infrastructure for emergent forms of life become apparent even as they refuse stable delineation (Lahsen 2010; Rudiak-Gould 2011).

Indeed, one of the impetuses for this research is based on an observation that climate change demands that analysts find ways to account for the "interdependent world" that Jay Rosen (1999) has described in relation to new forms of public and digital journalism and the richly textured terrains where "every action, twist, or turn of the lay of the land reorients all the other players" (Fischer 2003, 2). Fischer has argued that current modes of pedagogy and social theory aren't able to address fully the kinds of questions that emerge from techno-scientific problems. Such problems bring into focus "heterogeneity, differences, inequalities, competing discursive logics, and public memories; complex ethics of advocacy and complicity; and multiple interdigitated temporalities" (2003, 39). They require new modes of analysis and research methodologies that are multisited, multivoiced, multi-audience, and rework traditional comparisons, recognizing the traces and sedimentations of other analyses so that what results are structural critiques of the processes by which perception is being refigured through strands of sociopolitical networks (Deleuze and Guattari 1987; Dumit 2004a; Fortun 2003; Foucault 1995, 2003; Haraway 1991, 1996; Jasanoff 2004; Marcus 1995, 1998; Sunder-Rajan 2006).

This book seeks to provide both methodological innovation and structural critique on both these fronts. Specifically, it seeks to *adjoin* knowledges, taking into account relations of power and submerged and subjugated engagement with scientific facts operating in, around, and through the production of rationalizing discourse for mass publics. Climate change is examined as an emergent form of life in order that differences might come to light. As a method, this requires tracing specific communities, epistemologies, and related strands of translation, vernacular, expertise, concern, and collaboration. This book asks not only how individuals know what they know but how groups come to recognize the need to address climate change from their own ethical and moral reference points and establish other logics and baselines that come alongside the scientific evidence. It also tackles research questions in a way that holds in synchronicity the circulation of climate change reporting in the mainstream media and the particularities by which groups have adapted these scientific facts to their own vernaculars.

This book recognizes the need to think through ethnography as an account of *networked social relations*—considering both the technological and social aspects of such transformations. As Jane Singer has pointed out, "no single message is discrete" in a networked society, and "producers and consumers are not only interchangeable but also inextricably linked" (2007, 90). Network and information technologies are remaking lines of connection, radically effacing or remaking both flows of power and notions of cultural, historical, and geographical meaning (Benkler 2006; Boczkowski 2004; Castells 2000, 2005; Hardt and Negri 2000, 2004; Jenkins 2006a; Rheingold 2003; Tumber 2001; Turkle 1995; Turner 2006). While historians of media would point out that revolution has been predicted for every new form of media, current transformations related to the Internet present both opportunities and challenges for groups struggling to get their messages heard and actively produces new configurations of actors and experts.

In thinking through transformations of ethnographic tools and analysis that might capture networks and new configurations, Fortun has argued that in the face of double binds, for those who study technoscientific problems, "the actual is to be found in processes and intersections, rather than in objects and locales" (2001, 16). In so doing, locales come into sharp relief with one another, and as Fortun wisely articulates, what emerges is a partial view of the whole, recognizing that there is always more to address than is possible. It is this focus on processes and intersections that sets the work of this book apart from prescriptive analyses and reports that advise science experts, activists, or policymakers. The kinds of prescription that might emerge from the observations and analyses in this book address how to think about public engagement and the role of various kinds of media rather than offering a set of suggested directions. A focus on processes also sets this work apart from a work of journalism.

Prior to earning a PhD, I spent almost a decade working as a broadcast journalist and producer. In my academic career, a path I couldn't have predicted while undertaking the research for this book, I have chosen to teach in a graduate school of journalism. Returning to a mixed academic and professional environment has been an interesting journey. It has made the differences between ethnography and journalism that much more stark from my perspective. Journalism is about finding "the story" according to established rubrics within the profession and/or publications for which the story is intended. Editorial functions intervene at various points from assigning stories to devising headlines and paying attention to layout, placement, audience expectations, and competitive framing of the story. Some long forms of journalism are comparable in approach, depth, and time frame to anthropological methodologies.[19] Yet there will always remain key differences: the kinds of orienting questions being asked, relationships established with key interlocutors, perception of one's role as participant-observer, intention of publication, ethical oversight (academic work is overseen by independent review bodies whereas journalists rely on voluntary, unenforceable professional codes of ethics), and depth of analysis and situatedness (within larger disciplinary conversations) required for anthropological texts.

At first glance, the itinerant multisited fieldwork that informs this book may seem like a form of parachute anthropology (oxymoronic as such a term may sound to an anthropologist) akin to the critique of parachute journalism. Parachute journalism is a term used to refer to journalists who are dropped into a foreign country for a short period, often during times of upheaval when news events are taking place. As a 2011 article by Justin Martin in the *Columbia Journalism Review* succinctly put it, "'Parachutist' is a pejorative in the news business, based on the sense that an outside journalist who stays in a country or town for just a short time is unlikely to have a sufficient feel for the area's political and cultural landscape." Martin, a journalist and faculty member at Northwestern, goes on to argue that foreign reporters fail to adequately report when they are parachuted in, not because of the short time on the ground but because they haven't had the time or space to do enough background research. Deep context—historical, political, or otherwise—is often what's missing in media coverage of so-called hot spots, where conflicts and/or political upheaval are occurring. Why context is missing and how to address this industrywide challenge, particularly in these times of economic restructuring in the news industry, changing audiences, and new platforms, remains a much discussed subject among journalists and academics alike (Sambrook 2011; Singer et al. 2011).

But there is something deeper here to unpack than merely time spent on the ground. Instead, the notion of context, as I argue with climate change, might also be considered a form of life in which agreement over meaning is contested and differently situated in varied groups. Paul Rabinow in *Marking Time* (2008) compares journalism and anthropology by contrasting a journalistic sense of "subjectivity" with the academic commonplace of "contextualization." These two terms are used by both journalists and anthropologists, but are understood quite differently in their respective vernaculars. For journalists, the notion of bias or subjectivity retains an enormous amount of professional currency and with it an implicit nod toward professional journalistic ideas of objectivity (Schudson 2001, 2005; Singer 2005; 2007; Ward 2004). It is what drives the recognition of some experts over others, as well as accusations of exaggeration and alarmism. Chapter 2 wrestles with these aspects for science journalists reporting on climate change, but here I raise it as a way of differentiating the work this book seeks to do as a whole and what it does in contrast to a work of journalism. Rabinow considers anthropology and anthropologists' hopes for wider audiences beyond academic readers as somewhat "adjacent" to journalism—not unlike the concept of nearness threaded throughout this book in relation to advocacy and knowledge. For Rabinow, subject positions are defined *relationally*, fought for, and contain assumptions. Articulating these dynamics is the work of contextualization—work that falls far outside the terrain usually explored by journalists.

Locating one's subjects and objects of study, and oneself is the material work of contextualization that sets anthropology apart from journalism and renders the notion of bias both impotent and relevant all at once because it assumes we all have biases, positions, and subjectivity. There is no outside position or, to quote Haraway, no location that is "transcendent and clean" (1997, 36). Instead, locating oneself, objects and subjects of study, the techno-scientific, *and* the political is a texturing and textured process that relies on a "modest witness" stance, a stance based on what Haraway calls "strong objectivity." "Strong objectivity insists that both the objects and subjects of knowledge-making practices must be located" (37). Locating as a verb can be understood as recognizing the relations of power, knowledge, history, and difference. It is an acknowledgment of relations between things, institutions, regimes, people, and groups. Or, in more precise terms, it is about recognizing and articulating the strands of implicatedness of self as well as subjects and objects of study—the deep historical and social contexts that inform how things come into being, why some things matter, and who gets to define what matters.

Being "embedded relationally" or relationality is a kind of antidote to both relativism and transcendence, but as Rabinow points out, such an imperative recognizes both the constructed nature of one's object of study as well as the subject's and object's fluidity and evolution. In other words, one must choose what to study and how to study, keeping in mind that *worlds are made and unmade by what is considered significant.* Drawing on Clifford Geertz's observation, Rabinow notes that "anthropologists study villages, good anthropologists study processes in villages." But Rabinow underscores that this task does not arise spontaneously: "Choosing those processes is not mere witnessing but is itself an act of interpretation, or diagnosis." It is in this sense that itinerant fieldwork is perhaps more implicitly a felt set of choices where timing, location, witnessing, and participation are not givens related to residence but sought and juxtaposed at sites for observation.

The meetings for the groups I studied in this book didn't occur in only one location. Their multiple locations reflect the demands and requirements of participation in larger intergroup settings, activism, and/or disparately located group membership. Group and intergroup settings as well as semistructured interviews allow for insight into the worlds that climate change infiltrates, co-creates, and disrupts. They open up spaces so that articulations about what leaders and group members are doing might become apparent. Studying these groups demands a deep contextualization or strong objectivity such that how worlds are related, where and how double binds operate, and how it is that climate change is instantiated as a form of life comes into view. In order to navigate through these moving variables, Fortun suggests that anthropology draws from the experimentalism of the sciences, characterized by "an openness to what cannot be explained and to the possibility that what was once thought to be noise can be understood as signal" (2003, 186).

Experimental systems are reproductive, and they not only generate answers but provide orientation. They shape and clarify what the questions are and how they might be asked and answered. Experimental ethnography, as suggested by Fortun and Fischer and Marcus, similarly aims to construct an open system in order to rework and specify questions, examine and extend traditions of thought, and explore possibilities for shifts and displacements so that new insights and analyses might emerge. Resulting ethnographic texts are "like labyrinths, whose walls orient and limit the options available to a reader without telling her where to go." The result is still "a stabilized way of thinking about a particular phenomena," but what animates and differentiates such texts are the open pathways and discursive resources available for readers.

This research and the resulting book draw on this experimental approach such that the object of study, climate change, brings groups alongside one another and regards engagement with climate change as an open question. The itinerant nature of the fieldwork is not then a matter of parachuting *or* journalistic sensibilities. It is born of the necessity of following the object of study and, in the process, recursively and continually looking for clarification and understanding such that both questions and their answers come into view. This book narrates accounts of how this occurred in the field—where my questions to those I spoke with were completely reoriented, beginning with how or whether climate change is discussed and what it means in terms of the group or profession. Each chapter reflects a process by which climate change is instantiated within a group vernacular and thus represents a form of life. Layered together, they demonstrate how it is that climate change is understood as a scientific concept, an experienced phenomena, and a problem with moral and ethical contours that must be addressed personally and collectively in the American context. What it lays out is a path for similar observations in other national and global contexts and an experimental method that, to return to Wittgenstein, bring methodological challenges of examining public engagement with science and the role of media into sharper focus.

Structure of This Book

Chapter 1 begins in Alaska, where climate change is a lived and felt experience and where I first encountered a profound vernacular shift and resistance to the scientific and institutional definitions of climate change. It tracks the role of translation that leaders elected to the Inuit Circumpolar

Council representing Alaska, Canada, and Greenland endeavor to fulfill and the issues surrounding traditional knowledge, science, and media. ICC was formed in part as an answer to global governance needs, but it has emerged as a key interlocutor for integrating traditional knowledge and scientific knowledge *and* articulating Inuit self-determination in national contexts. This chapter asks questions about knowledge and expertise that will be addressed throughout the book.

Chapter 2 tacks back to wider efforts by media to engage the public, and it seeks to understand how science journalists deal with the challenge of articulating risks and scientific findings related to climate change. Adjudicating expertise is central to the practice of journalism, and climate change provides an evolving challenge. The analyses in this chapter take into account that journalists are now working in an atmosphere of shifting professional norms and immense technological transformations in media platforms. The differences between national and local media come to the fore in this respect, as does what has been afforded by new technologies. A central aspect of the challenges facing science journalists is how to articulate the message of climate change in ways that avoid alarmism and yet indicate an ethical need to act while they are simultaneously negotiating rapidly shifting professional norms, audience expectations, new participatory cultural forms, and emergent ethical frameworks.

Chapter 3 provides a distinct contrast to the work of journalism by looking at the rise of Creation Care, a new submovement within the evangelical movement. Creation Care seeks to shift evangelical Christians' view of the environment and make it a Christian priority and responsibility. Leaders within the movement regard who is speaking about climate change as a key to "blessing the facts." In other words, science doesn't speak with the same authority everywhere, but instead operates with a distinct vernacular that has been naturalized by some and rejected by others.

Chapter 4 shifts back to mainstream and scientific discourse, examining the work of scientists who have deployed their expertise in the high-profile debates and media coverage about climate change. Many journalists decry the inability of scientists to communicate their research adequately, while scientists are still reluctant to look like they are chasing media or public attention. The problem of advocacy set against the norms and values of their profession is something journalists and scientists share, and this chapter more fully investigates risk and insurance metaphors that are part of the ethical calls by science experts to address climate change. Chapter 4 argues that scientists are engaged in what can be described as a wide spectrum of near-advocacy in policy and media arenas where representations and vernaculars are continually jostling and interacting with one another.

Chapter 5 takes a closer look at what climate change means inside a vernacular where risk is the driver and incentive for financial institutions to respond to climate change. Ceres, a corporate social responsibility group, operates on several levels offering a stakeholder process for member companies and an Investor Network on Climate Risk. By translating and transforming the problem of climate change into one of climate risk, Ceres brings to bear the vernacular of business on a problem traditionally considered irrelevant to the proverbial bottom line. Climate change's form of life in this sense is one that has distinct parameters for measurement and management within a business context.

The book closes with an epilogue that is meant to underscore both the continually evolving nature that a study of public engagement with climate change entails and the ways in which its form of life is continually up for definition and redefinition. Rather than offering conclusions, the epilogue reflects on questions of media, democracy, and expertise that operate on shifting ethical and moral terrains, and the continually evolving definitions of what climate change means. It also returns to a consideration of experimental methods, refining how to think about public engagement—and how climate change comes to matter.

CHAPTER
ONE

The Inuit Gift

On July 7, 2007, I awoke early to a brilliant Arctic sun
already high above my hotel in Kotzebue, Alaska. Kotzebue is a town that guidebooks refer to as a "working
Arctic town," or what I determined as code for "nothing to see here." Such a description is in stark contrast
to nearby Nome, which caters to tourists, Iditarod
sledding enthusiasts, and gold rush history seekers.
I traveled the extra leg to Kotzebue so I could attend
the Inuit Circumpolar Youth Council (ICYC) language
symposium. The invitation had been extended to me
by Nome-born Patricia Cochran, international chair of
the Inuit Circumpolar Council (ICC).[1] ICC represents
Inuit people across the Arctic parts of Alaska, Canada,
Russia, and Greenland.[2] ICC has both a youth council
and an elders' council in addition to the main political
organization.

It was a privilege to be invited to the ICYC symposium in Kotzebue, but after I accepted, I realized the

symposium fell on the vaunted 7–7–7 date. I had originally planned to attend one of the Live Earth mega-concerts scheduled for that day. Live Earth, at that time, was one of the largest (and most expensive) efforts at generating public awareness and engagement with climate change. Many of the world's most popular musicians had signed on, and Al Gore's organization was programming the climate-related part of the program. It was meant to energize the faithful and convince others to care and do *something*—even switching light bulbs from incandescent to longer life compact fluorescents (CFLs) counted as a responsible response to climate change.

Each morning I was in Kotzebue, I would descend the stairs to the hotel lobby where a small group of male elders were chatting and laughing with one another in the seating area in front of registration. Tied together through networks of kinship and friendship, they came from various fly-in communities, like Point Hope, Kobuk, Barrow, and other villages in the northwest Arctic. The symposium was a reunion of sorts for everyone who attended. I was a bit of an anomaly, although they were certainly accustomed to scientists, social and otherwise, being in their midst to study them or their land.

The same group of elders had questioned me a day earlier about my identity. They were sure that I was a lawyer and had a good laugh when they found out I was a graduate student. Climate change as my topic of interest elicited a different response—the tone of the conversation shifted quickly. Several spoke very briefly and gravely of storms that had forced their whaling boats back in, changed game patterns, and continued dangerous erosion of their coastal villages. They didn't necessarily want to know what I was up to in an in-depth way, but they did want to inform me that these changes were very much an everyday concern for them.

On July 7, they were deep in conversation in their Iñupiaq dialect. We exchanged waves, and I headed out the front door beside them to be greeted by the gloriously bright sun and gently lapping waves of the Chukchi Sea. The dirt ring road about six feet from shore lay in front of the hotel and provided an easy footpath to the restaurant next door—one of only two or three places to eat out in a town of about 3,000 people. As I slid into a chair at the restaurant, I wondered if anyone in Kotzebue was aware or excited about the fact that somewhere in the world really famous musicians were rocking out about climate change to save the Arctic and, if one believed the most alarming projections, countries and land masses as we currently know them.

CNN was on in the restaurant, which doubled as a bar. It had updates from concerts under way in Tokyo and London. CNN's anchors were quite excited about the scientists' band broadcasting later on from Antarctica—excited, that is, in the canned performative way viewers have come to expect from on-air banter. I had to agree with their canned excitement, though. The "broadcasting from all seven continents" was a real novelty even if the seventh came by way of grainy satellite video from a socked-in Antarctic winter research station. That was it for a human polar presence, though—from the only continent devoid of indigenous human communities.

I glanced around the gritty restaurant with faded leather chairs and paneled walls. It occurred to me pretty quickly that I was the only one paying attention to the screens mounted on the ceiling above the bar. The wizened old fishermen in the booth behind me were talking about the relative merits of various winches and rigs. The elder Inuit couple and their grandchild in the booth beside them talked quietly. I couldn't make out what they were talking about, but they gave me a gentle nod to say hello, recognizing me from the symposium. Other breakfast-seekers straggled in over the next forty-five minutes, but the TV was mere background noise. Game day or election night this was not.

The Arctic was not center stage for Live Earth, despite the daily challenges of living in a vast expanse dotted with fly-in communities that have worked out a dependent relationship with ice and cold. The irrelevance of such an event to those actually experiencing the direct effects of climate change seemed palpable from this vantage point. Learning about compact fluorescent light bulbs just doesn't cut it as a solution when nearby, the ancient whaling village of Kivalina is in danger of being swept into the sea or, to put it less dramatically and more specifically, losing more and more of its small barrier island to permafrost melt and coastal erosion.

It has been argued that awareness-raising schemes like the massive undertaking of Live Earth are always removed, regardless of where one sits. Certainly, there were many critics and skeptics who wondered what the "real" net effect would be in terms of greenhouse gas emissions and the expense of broadcasting musicians like Sting, Madonna, and the Black-Eyed Peas live from large and fashionable metropolises like Tokyo, London, or Rio (Schagen 2007). Yet for those who long for a continued momentum of public interest and support for climate change action and the energizing of a new generation, there could be nothing better than a Live Aid for the Earth. After so many decades in which climate change remained on what Gallup called "the public's 'back burner,'" it finally seemed that such a massive event might be a way to raise the profile of climate change the way Live Aid or Farm Aid or other celebrity-laden events had done for other issues.

Between this gulf of the local and global, the direct present experience and the conceptual future, lies the difficulty of communicating the amorphous nature of climate change as an issue of concern. How to talk about it, where it's located, what the causal factors might be, when it may begin or how it already has, and any guesses at potential solutions appear, at first glance, to be audience-dependent. Locating what climate change means or when/how it is meaningful is a much more fraught process than what advocates or journalists might consider in their efforts to make an issue or a news story relevant to "audiences." The rules, grammars, and associations related to climate change's form of life are in motion. Locating oneself or an event in relation to climate change brings to bear history, collective identities, institutional regimes, and epistemological difference. For example, the fact that Live Earth tries to connect climate change not to where it's happening or the people located there but to musicians and science and policy experts invests it with certain kinds of knowledge and politics.

I am beginning this book with my journeys to the Arctic because this is where climate change is a lived and felt reality. Beginning where climate change is already happening reveals the ways in which climate change has been shaped by scientific vernaculars and media discourse and the ways in which its form of life requires many different levels of negotiation by those who are implicated in the predictions and experiences associated with it. This chapter seeks to locate climate change in diverse Inuit discourse, contexts, and histories, and to locate Inuit claims and experiences in media and science-laden contexts where action, logics, and representations compete for dominance and prominence. Understood as an emergent form of life, particularly in the Arctic, climate change presents the need for excavation and reassessment of what a recognition of climate change portends for those who have endured a century of immense cultural, political, and environmental changes. My stark awakening to this began a day earlier, before I went looking for global climate change concerts.

Climate Change as "a Three-Month Conversation"

When I first flew into Kotzebue, I wasn't sure what to expect. Most of the travelers at that time of day seemed to be local people. Though Kotzebue is small, there were two or three taxis waiting for disembarking passengers. I surmised that either locals often needed transport or much more traffic than I was aware of passed through here on the way to the nearby Red Dog mine (90 kilometers away), the offices for the Iñupiat-owned Nana Regional Corporation, or any of the villages that formed a hub around "Kotz."

My taxi driver wanted to know where I was from, what I was doing in town, and if I had ever been this far north. The thing about small towns is that once you land, there is a sense of obligation to identify and locate oneself among the pantheon of previous and future visitors. The ride to the hotel was probably about three to five minutes, yet it seemed to last much longer, providing me with my first glimpses of the Chukchi Sea.

The woman at the hotel desk was considerably more professional and urbane than the taxi driver. She checked me in without any small talk and gave me directions to the school gymnasium where the symposium was taking place. It was about a ten-minute walk and my first opportunity to wander through town. The streets were mostly empty and lined with weather-worn wooden houses and buildings. The Quaker church, a bright red barnlike structure, jumped out at me, as did the array of large satellite dishes, which I later figured out were next to the building that housed the radio station. Further from town, one could make out reddish-orange cranes at the shipping terminal on the edge of the water—the familiar outline of a working seaport.

I wandered down the gravel road that was a main street through town and found the school with little trouble. I entered what seemed to be the front door and followed the sounds of voices down the hall and past a large trophy case. There was no signage denoting the symposium, and hardly any hall lights were on.

When I entered the brightly lit gym, I had the distinct sensation of joining a community meeting of eighty to one hundred Inuit people. They mingled, drank coffee, and gossiped. The only people I vaguely knew were Patricia Cochran, whom I had met briefly at Arctic Science Summit Week some months earlier at Dartmouth College in New Hampshire, and Megan Alvanna-Stimpfle, chair of ICYC, whom I had spoken with on the phone to secure her agreement for my attending the symposium. Needless to say, I was a bit of a curiosity. There were two other non-Inuit social science researchers there, I discovered later, but they were well known to all the participants, having either lived in Kotzebue or worked with the youth for some years.

I was greeted warmly. Most of the elders and leaders made a point of finding out who I was. Some thought I must be Inuit, which I'm not. I explained to many that I was an enrolled member of what was, for them, a "southern" Canadian tribe (the Tahltan Nation located near the Yukon/British Columbia border), but I was studying at MIT in Boston, and my research looked at the communication of climate change to Americans. My MIT status garnered more interest than my indigeneity, especially because I wasn't part of the sociopolitical or kinship fabric of any Arctic or sub-Arctic indigenous group. The responses to my research topic were varied, but one of my first conversations was transformative in a way I didn't anticipate.

A prominent locally elected official, upon hearing my personal and research introduction, said, "Climate change . . . we don't really talk much about that. It's more something they talk about on CNN. It's out there. It's not what *we* talk about." I was shocked by her comment, but intrigued as well. I wondered if I had misunderstood either her comment or climate change in the Arctic because environmental change related to massive warming trends all over the Arctic were being heavily discussed in Alaska at that time. The headline on the regional paper for the Kotzebue area announced the recent hearings by the Alaska Climate Impact Assessment Commission, which had two individuals from Kotzebue on it—another elected representative and an elder. Kotzebue is also a hub for ten nearby villages accessible mostly by boat, one of which is Kivalina.

Kivalina makes for a striking visual image. It is thousands of years old, located on a tiny barrier island whose edges are slowly being reclaimed by the sea. Several months after my trip to the area, the village leaders filed a lawsuit against major oil companies in order to cover the cost of moving their village from its barrier island to another location. Shishmaref, further north, had also been in global news reports on climate change—what some took to calling "climate porn" (Lowe 2006). Shishmaref can represent climate change in a way that makes it "real" and horrific by showing houses and a shoreline destroyed by permafrost melt, coastal erosion, and changing weather patterns—all attributable to climate change. In fact, Cochran had told me that earlier that year, Shishmaref had asked her to pass on the message to interested media to give them a break for a bit. They were so inundated with media crews that it was beginning to become a problem for the small village. Climate change then was recognized as a serious challenge facing residents of Arctic Alaska *now*, not sometime in the distant future, and conversations with others revealed this quite vividly. How then to make sense of the resistance to climate change?

Hours later, I talked with Patricia Cochran about the CNN comment. She squeezed in a longer interview with me between conference sessions. We sat on a bench in front of the hotel, facing the Chukchi Sea—its waves gently lapping about three feet from us. Every so often, old friends or conference attendees driving or walking by would stop to say hello to her. We watched as a boat filled with younger men pulled out for points across the inlet. It was a beautiful view, with the sun high in the sky and the inlet seemingly going on forever in all directions.

It's not that people don't talk about climate change, or are unsavvy about the term, Cochran told me. They just don't necessarily call it that. Rather, the everyday vernacular in Kotzebue and among those from other communities throughout the Alaskan Arctic tends to focus on symptomatic changes along the lines of the elders I earlier described chatting to in the hotel lobby—whalers forced back in, more storms, more intense storms, early sea ice breakup, and coastal erosion.[3]

> Certainly, when our elders talk about climate change and global warming, those are not the words that anybody would ever hear coming from an elder's mouth or anybody else. Maybe because those are just not the words that we use. But if you were to ask elders about the changes in ice conditions, and what they have seen in their lifetime, changes in ice? Well, that would be a three-month conversation.

The absurdity of trying to sum up a lifetime of discrete observations layered on oral histories and community consensus about witnessing environmental change in one term is striking, particularly for those who have a tendency to gloss over the definition of climate change as something to be found in the pages of *Science*.

And yet is this "three-month conversation" the same as "climate change"? What does a rejection of climate change in the place where it is seen to be happening mean? What kind of problem is it that *climate change* isn't a recognizable term, and one assigned to media (CNN in this case) as "their" term, as something experts and journalists talk about and not what's happening in and around this part of Arctic Alaska?

When Fischer expands on the social life of language and knowledge offered by Wittgenstein, he argues that techno-scientific problems present as emergent forms of life, bringing to the fore a direct confrontation with "the other," with heterogeneity and historical genealogies. I wondered what genealogy lay behind this difference and differentiating between "them" on CNN and "us" and the refusal to defer to climate change. And I wondered at this occurring at a moment in which certainty seemed to have been achieved in the wider science and science policy world—where terms like *scientific consensus* and images of Shishmaref were regularly trotted out as evidence of some kind of closure. Wittgenstein describes certainty in language as the point at which questions no longer need to be asked and explanations come to an end. Yet here I experienced not an explicit questioning of climate change but a flat-out rejection of it as a term that described what direct experience with climatic changes feels like and how it is that such changes are understood and discussed.

Cochran expanded on the notion of a three-month conversation by weaving vernacular and worldview together.

> It has a lot to do with different language. I don't mean different Native languages, but the way we use common everyday language. And then the other piece of that is the Native worldview. All things are connected, and so to take one piece of a problem and not connect it to the rest of the world and the environment around? It just logically makes no sense. How can we talk about changes in weather without talking about changes in vegetation or the air or the people or the animals, as all of those things are part of a natural mix. All things are connected in our universe.

The point Cochran makes here is that the ways things get talked about have a direct relation to one's experience and point of view. Or perhaps more succinctly: how one talks about the environment is based on how one comes to know it. In academic settings, our word for referencing or studying how we know what we know is epistemology. It references the ways in which knowledge has a history and a genealogy and the ways in which how we both know and speak about what we know are situated within cultural learning and contexts.[4]

For many Inuit people across the Arctic, this learning process involves communal and familial interactions with elder family members and time spent on the land. Hunting, fishing, and whaling traditions differ within communities, and all Inuit do not have the same opportunities to learn.

The transfer of tacit and formal knowledge in any context is not an even or automatic process, and in the Arctic, histories matter and vary between villages, regions, and countries. It wasn't until the 1950s that many of the assimilation practices practiced elsewhere in the U.S. and Canada reached the Arctic, where children were shuffled off to schools and previously nomadic indigenous people were forced into structures and stable locations.[5] For many, this was a violent and destructive process, marked by imposed colonial regimes, relocation, cultural change, and resilience. In part as a direct result of this colonial legacy, Inuit communities also suffer from many of the same social and economic challenges that plague small and rural communities across the north and far north. Suicide prevention was a primary concern of the Inuit Circumpolar Youth Council's international council, in addition to culture and language retention. Working in the social and political world of ICYC, Cochran, and many others associated with ICC involves a continual confrontation with and response to change and the historical sedimentations of diverse colonial policies of assimilation and geopolitics in the Arctic. Climate change is a recent newcomer to this discussion and the attendant negotiations over what should be prioritized.

One of the moments that made this evident was a meeting I witnessed between Cochran and ICYC council members, all of whom were in their early to mid-twenties. ICYC's council was a diverse and accomplished bunch. Some were already parents, and many were recognized community youth leaders who had worked on issues related to language and community survival. The council, being newly formed, was using the get-together as an opportunity to map out priorities for the coming four-year term. There were two representatives each from Canada and Greenland and several from Alaska, including the chair and vice chair. This meeting in Kotzebue was one of the few in which they would see each other face-to-face. There are no direct flights across the Arctic, and northern travel in general is extremely expensive. Megan Alvanna-Stimpfle, the chair at that time, told me that they usually used social networking tools, voice-over Internet protocol applications like Skype, and e-mail to communicate.[6]

Cochran had come to the ICYC meeting in order to talk with the ICYC leadership about what ICC as a whole was doing at the policy level where Cochran and other ICC leaders work and to talk about climate change. ICC provides a political platform as well as lobbying opportunities in global and domestic political regimes so that Inuit needs and priorities are represented and so that governments and their policies might be held accountable for the decisions they make that affect Inuit people across the Arctic. ICC was formed as a response to industrialization and militarization of the Arctic and the urgent need for indigenous representation in both national and transnational policy arenas. ICC has an international office with a rotating chair from each country for a four-year term. ICC then has domestic offices in each Arctic country with their own priorities. What became clear to me at the various events where I interacted with ICC representatives is that leaders like Cochran work simultaneously at the village level, where direct observation, experience, and Inuit traditional knowledge are the infrastructure for one kind of civic epistemology and form of life, and at the transnational governance and media level, where, conversely, scientific findings form the basis for a vernacular to discuss and describe climate change.

ICYC members are less likely to take the lead in tacking back and forth between these very different worlds, and their difficulty was in reconciling the more pressing needs like language retention and suicide prevention against an issue like climate change. In some ways, they represent a microcosm of the challenge presented to national and transnational indigenous organizations where multivocality, regional priorities, and negotiating a balance between various social needs means accepting trade-offs of one kind or another. ICYC leaders discussed climate change in these terms at some length, seeking to situate it among (a) what they could contribute to the discussion, (b) where and how it fit in their national contexts, and (c) how to reconcile it against what they know to be the pressing needs of youth across the Arctic. They didn't come to a conclusion as to where climate change would fit in their pantheon of goals, but the prospects of climate change being a central or top priority did not seem likely given the more pressing needs around suicide prevention and language retention.

When I spoke to representatives from Canada and Greenland afterward, they both mentioned that they had spoken to media about climate change and had seen their role in ICYC as needing to fit with the priorities of both ICYC and their country's ICC council. ICC Canada and ICC Greenland had made climate change a high priority, and it was a natural fit that the youth would speak about this issue in their home countries. ICYC priorities were thus meant to guide their work as a council, but individuals were free to speak as they saw fit within their national contexts.

Within national contexts, climate change has different meanings and political baggage of its own as do indigenous voices, leaders' and organizational aspirations, and the needs of communities. Certainly the American/

Alaskan context presented a very differently configured challenge than the Canadian or Greenlandic one at that time. Canada and Greenland were both signatories to the Kyoto Protocol and actively working at the national and transnational levels to reduce and mitigate for emissions.[7] As well, Greenland is in the latter stages of a transition to Homeland Rule, an innovative thirty-year-long structured transfer to self-governance from Danish rule. Canada has chosen a different path, but one that ensures a high profile within the Canadian political and media context, marked by major events in the 1990s such as the settlement of Inuit land claims and the creation of Nunavut, a new Inuit-governed territory. In the United States, however, Iñupiat, Yupik, and Cupik people were part of the Alaska Native Claims Settlement Act (ANCSA) signed in 1971 and were organized into regional corporations. Russia is a much more complex situation where indigenous rights are not legally or formally recognized through settlement or other agreements.

I spoke with Sheila Watt-Cloutier, ICC Canada's former chair and Cochran's predecessor as the international chair, and described the difficulty I had witnessed about prioritizing climate change. She responded by articulating her rationale and means for prioritizing this way:

> I think that some people have not fully come to understand that there is no disconnect between the suicide rates in our communities and climate change. There is no disconnect there. Environmental issues— it's all connected. I don't know what Alaska is like. I cannot speak for Alaska. But I know that many of our young people remain quite connected to a hunting way of life. If they don't, then their parents do. There is a real connection going on with the way of life and yet even with that I remember getting a question a couple of years ago. Why do you focus so much on environment and not social issues at the ICC level? I said there is no difference between the two. I mean, it's all connected. You have to look at the larger picture of how our hunting culture is not just about going out and killing animals; it is about preparing our young people for everything, challenges and opportunities. And it is because of that disconnect between our children being prepared with the character building that a hunting culture gives and the institution separating that completely in terms of how to be taught, how to be patient, to be bold under pressure, to withstand stress, how to be courageous, how not to be impulsive, how to have sound judgment and wisdom. That is all the hunting culture that gives that.

In Watt-Cloutier's formulation, climate change continues the process of foreclosure on hope, begun by encounters with colonialism and the enduring structures it put in place via education and mechanisms (or lack thereof in previous eras) for governance, communication, accountability, and self-determination. The environment is also an extension of and constituent to culture. The response from Watt-Cloutier as well as the ICYC meeting and Cochran's formulation illustrate poignantly the multivocality within the Inuit communities that ICC represents (Steinberg 1999; Terdiman 1990). It also explains the process of relationship building with the facts that I referenced in the introduction. This process of articulation and translation must first occur within a group in dozens of small and large conversations like the ones I witnessed before it can wend its way out into positional articulations that put meaning, ethics, and morality front and center for media and wider publics.

It's not only a relationship-building exercise with the facts that is at stake here. There are also the attendant institutional structures, priorities, and categories that have sprung up around the instantiation of climate change as scientific fact. Wittgenstein claimed that training is required such that a concept becomes shared collectively and its form of life emerges. We know what something is and what it means through a system of rewards that occurs in the learning of a language in childhood. Using such a framework, one might argue that a conversation about prioritizing is actually a conversation about accepting climate change as a shared fact and one with political stakes. Conversely, resistance to prioritizing climate change might also be seen as an extension of the resistance to integrating it into everyday discourse about observable changes. Part of the resistance to climate change as "something on CNN" thus might be conceived as a resistance to what climate change *imposes*. The gulf then is not between the local and global but rather between the symptoms/experience of climate change and "climate change" as cliché.

French philosopher Gilles Deleuze describes cliché as something that creates comfort, doesn't allow one to wake up to the intolerability of the present, and lacks the descriptive or depictional powers that might allow real change to occur (Deleuze 1989; Dumit 2004b). Climate change gets defined in one way in the science, policy, and media discourse that teeters continuously on the edge of cliché but it gets defined in a completely other way "on the ground," where it's happening in real time. This is part of what makes it difficult for those experiencing it to recognize it as their own when it shows up, for example, on CNN, and it's also what makes it difficult for reporters to cover it as an in-motion still-being-defined form of life that requires action. One of the enduring frustrations for advocates and near-advocates as I will detail in chapters 2 and 4 is the challenge of how to reveal facts so that audiences will wake up to the problem that climate change–related predictions present and the need for action, both personal and political.

These multiple definitions and ways of understanding climate change are as much about timing and distance as they are about the institutions and processes that produce a point of view and an overriding "correct" definition. Defining climate change requires a translation process that is one part transformation through assigning meaning in order for the concept to take root in other vernaculars, and one part training in order to recognize the institutional challenges and opportunities that arise through aligning one's experience with "climate change." So for Inuit, this "thing" that is happening, that is noticeable and felt—changes in the ice and other indications that all is not as it was—got described differently, more specifically long before the term *climate change* came to take its place. And one might then surmise that when "climate change" did or does arrive, it comes with its own sets of baggage—or rules, grammars, exceptions, and associations in Wittgenstein's terms—that like any invading army is not greeted with the kind of embrace scientists or journalists or researchers like myself might expect. This is an issue not of complete incommensurability but of epistemological difference and a confrontation with history, institutions, and power relations.

What such a framework puts on the table is a key question: Is the refusal of climate change that I first encountered a moment in which climate change is lost in translation, a complete rejection of the system that climate change represents, or one in which climate change hasn't yet become a means for describing the Inuit experience with climatic changes, scientific research, and/or geopolitics and transnational institutions intent on reordering worlds according to this newly agreed upon fact? If the latter, translation might thus be considered a process of elaborating climate change's form of life, as well as recategorizing Inuit experience and negotiating on a new terrain with many of the usual suspects: science, media, transnational institutions, and national governments intent on pursuing what is currently perceived as being in their "national interests." Such a translational process is never frictionless or homogeneous, and it allows for multivocality and multiple interpretations. It reflects the emergent aspects of a form of life yet to achieve anything like certainty or commensurability and for which explanations continue to be needed and negotiated.

I want to first explore what this kind of translation looks like on the terrain of science and scientific claims about climate change in the Arctic. I've only hinted so far at the epistemological differences behind a resistance to climate change, but this next section addresses what's referred to as "traditional knowledge" and locates Inuit experience and knowledge within and alongside scientific facts, experts, and institutions related to both science and policy. Exploring this terrain sets the stage for the following sections that seek to contextualize and understand efforts by key individuals like Sheila Watt-Cloutier who have sought to instantiate Inuit experience in climate change and media discourse by "putting a human face" on the issue and interpreting environmental changes in the Arctic within a human rights framework.

Out on the Tundra and Beyond:
Ground Truthing versus Model Truths

When I went to Kotzebue, I didn't come to speak to elders specifically as most scientists and researchers usually do when they're working on climate change. But I did end up with the distinct privilege of having an elder take me out on the tundra to show me firsthand the signs he has observed regarding climate change. In fact, in a departure from many, when I told him about my research, unlike everyone else, he was deeply interested. Caleb Pungowiyi is one of two Kotzebue area residents who sat on the Alaska Climate Impact Assessment Commission, which held hearings in communities around Alaska about the impacts of the changing climate (2008). In addition, he was also an Alaskan chair of the Inuit Circumpolar Council in the past and had worked on numerous governmental committees.

Pungowiyi said he is often asked to sit on such committees, conduct talks, and partner with researchers not because he is "the most knowledgeable" but because he has "that way of communicating to the learned community about what's going on"—he has the ability to observe and talk about it in a way people understand. The implicit underlying point of Pungowiyi's is similar to Cochran's point—that the way climatic conditions are discussed in the village differ starkly from scientific, policy, and other kinds of vernaculars.

After one of the afternoon conference sessions, I traveled out on the tundra outside of Kotzebue with Pungowiyi in his small pickup truck. It was a gorgeous sunny day and my first experience of visually interacting with the Arctic landscape outside of the town. He pulled over at various points and explained various markers of climate change.

Pungowiyi began by showing me fields of cotton grass that had moved in. Cotton grass is beautiful, dotting the landscape with swaying low-lying grass, much like prairie grass, but with a cotton puff on the end. He showed me evidence of moose markings—moose are generally found much further south in plentiful numbers. He also showed me evidence of melting permafrost, picking up chunks of soil to show me how dry it had become. He said that more moss is growing, pushing out the lichen that caribou feed on. Caribou are a dietary staple in this area and throughout the Arctic and sub-Arctic.

I asked him if sea and sea ice changes were equally evident to his trained eye. He said that, though he was not a whaler, it was well known that small boats had a hard time hunting when the wind picks up. And he said that changes in wind, water temperature, and precipitation had produced all kinds of changes that scientific instruments miss.

He said he had started to notice the changes twenty or thirty years ago. He originally thought it was weather variability and that it was a blip that would change back. But the moose didn't move south again; they kept coming, and they stayed. Things that were surprising indicators of change decades earlier had now become more common. In other words, it took a keen eye observing over a longer period of time to recognize how much the landscape was providing signs of change and what that change was.

Unlike the treeline in the south, the expanse up on the Arctic tundra is usually wide open—the land intersecting with the horizon in the distance. Kotzebue used to have one black spruce tree—an oddity on the tundra. Pungowiyi said that a couple guys had hung a sign on it as a joke that said "Kotzebue National Forest." But owing to climatic changes, a few more black spruce trees had appeared in recent years. The trees are kind of a harbinger of change—change that will have an effect on life throughout the entire Arctic. The sign came off some time ago. It's just not as funny as it used to be.

While I tried to take in the details and grasp the weight of these small but steady changes on the tundra, what struck me is this: in the midst of rapid urbanization and the past hundred years of industrialized special-ization and detachment from varied connections to the outdoor environs, climate change requires individuals to have grounded knowledge about the natural world. But it also requires an ecosystem mind-set—a capacity to develop a mental framework in which bits of information can be dynamically plugged, reworked, and seen to interact. For the overall picture, Pungowiyi said they look to science, but they depend on their own observations for *ground truthing* that scientific instruments miss.

As it is practiced in various subfields and disciplines and as a methodology, science is another knowledge system, another mode of apprehending and deriving predictions and patterns concerning the natural physical environment. What is striking is the tacit understanding that the language and views espoused through the practice of these scientific disciplines drive mainstream media dialogue as well—a point scientists frustrated with media representations of their work might well dispute and which I'll address more fully in chapters 2 and 4. That scientific methodology governs the way the natural world is measured, ordered, and understood when the environment is covered by American media is taken for granted. Indeed, the language of the sciences is the default common vernacular for mainstream western society when it considers the environment.

In contrast, what Cochran and Pungowiyi describe is referred to as "traditional knowledge" (TK), and I would argue that it remains well off the radar in U.S. public fora, except in Alaska and when controversies erupt of the legal or other variety. Kivalina provides one such example. The well-regarded nonacademic book *The Whale and the Supercomputer* by Charles Wohlforth (2004), which profiles figures in TK, science, and science policy in tandem, also move such distinctions more concretely into popular media representations. But, arguably, the biggest move to bring traditional knowledge into constructive relations with various scientific fields is in the Arctic Climate Impact Assessment (ACIA).

ACIA was released in 2004 by the Arctic Council, and it broke new ground in its billing as a thorough combination of traditional and scientific knowledge. The report heavily involved Inuit TK experts and more than three hundred scientists. What wound its way out into international news was the remarkable disparity between predictions for the poles and the rest of the globe: where the southern United States might feel a two degree rise in temperature, the Arctic would be looking at a change of ten degrees or more. The sensitivity of the Arctic to such changing temperatures and its subsequent cascading effects for the rest of the globe in the form of melting glaciers, sea ice, and correlative sea-level rise made propositions of Arctic change highly relevant for worldwide consideration.

When Watt-Cloutier announced the Inuit human rights petition, she used the ACIA as a key point of evidence, but not just in terms of its findings. Her references to ACIA focused equally on the process by which the findings were arrived at: "We know that science and traditional knowledge are saying the same thing. What we have been saying for years now, science is affirming, confirming." On one hand such a statement can be seen as a tiebreaker in the fierce scientific debates that were coming from industry and the Bush administration at that time: Inuit oral history and knowledge of the sea and ice add irrefutable tacit evidence to the mounting "consensus" among scientists. But what Watt-Cloutier's rhetorical move also does is tie the Arctic region to Arctic peoples and specifically to the Inuit.

During the past decade, scientists, environmentalists, and journalists have routinely referred to the Arctic with monikers like "canary in the coal mine," a "world health barometer," "bellwethers for all of us on planet Earth," an "early warning system," a "sentinel," or a host of other descriptors and metaphors evoking a fragile, affected ecosystem metonymic of the earth's fragility as a whole.[8] Yet as remarkable and natural as all this seems given the current scope and predictions related to climate change, the focus on the Arctic is a rather recent turn. As Watt-Cloutier has stated in numerous venues, the Arctic was not even mentioned as a vulnerable area in the 1992 text of the United Nations Framework Convention on Climate Change. Instead, "low-lying and other small island countries, countries with low-lying coastal, arid, and semi-arid areas or areas liable to floods, drought and desertification, and developing countries with fragile mountainous ecosystems" were named as particularly vulnerable. The omission of the Arctic or Arctic nations as vulnerable is particularly poignant considering its transformation into an increasingly ice-free area directly affects sea level rise and flooding of those low-lying areas. ACIA clearly and permanently corrected this oversight.

Conceptualizing the Arctic as ecological and vulnerable may counteract the UN's original framing of what parts of the world are most "vulnerable" to climate change, but it still leaves the Arctic people a distant, if at all visible, recipient of the effects of climate change. Megafauna, like the polar bear, are still more likely to make the cover of *Time* magazine. Yet if we follow Watt-Cloutier's formulation, the knowledge of the Arctic exists in relation to its inhabitants. Science is the interloper, providing a translation, legitimation, and other language of expression for the rest of the world. It's in this sense that ACIA represents a hybridity of expertise and marks a major milestone.

It's only since the 1980s that terms like *TK*, *traditional ecological knowl-edge* (TEK), or *indigenous knowledge* (IK) have been widely used, and then often only in indigenous, academic, or policy arenas. The concept that TK embodies has its roots in anthropological and explorer recordings of travels beyond western civilization. Most, if not all, such travelers to the Arctic have expressed amazement at the deep local knowledge of climate, ocean, land, plants, and animals. And often their lives depended on such local knowledge in order to survive harsh and unpredictable conditions. These records encompass what Claude Lévi-Strauss (1966) theorized as "the science of the concrete"—the search for order in nonwestern civili-zations, which is not primitive in the sense of an evolutionary step that precedes an enlightenment through rationality and science. Rather, it is, as Cochran notes, a separate knowledge system that is sometimes said to have its own evolutionary path of development. Such a path, it is easily argued, has been in a state of détente with science since the arrival of colo-nialism, despite providing science with essential insights and methods at crucial historical moments (Cruikshank 2005; Fischer 2003, 2009; Grove 1996; Koerner 1999; Wohlforth 2004).

TK is defined variously as qualitative, intuitive, holistic, moral, spiri-tual, empirical, lived, oral, systematic, detailed, and diachronic as opposed to the specialized, quantitative, rational, synchronic, systematic, detailed, objective qualities usually associated with science. But the line between scientific and traditional knowledge is less stark than such a laundry list would suggest, particularly as varied models of collaboration have begun to emerge, but even in previous decades, TK could prove enigmatically useful. TK was instrumental in the 1970s in supporting the claims of Inu-piat whalers in Alaska who successfully challenged scientific data related to the migratory bowhead whale population (Benson 2008; Berkes 1977; Bodenhorn 2003; Feit 1987; Inglis 1993; Wohlforth 2004). Scientists even-tually were forced to agree with their TK-based calculations and observa-tions, and the moratorium on subsistence whaling by the International Whaling Commission (IWC) was lifted as a result. There are similar exam-ples where diachronic and quantitative data have been generated through TK, but what makes it entirely separate from the project of science and the scientific process is its social context and production.

TK is part of a worldview that generates symbolic meaning from the environment, which is shared communally and historically. As anthropol-ogist Julie Cruikshank (1991, 2001, 2005) reminds us in her investigations of local knowledge and glaciers, TK is porous and socially situated. Harkening back to Cochran's comments, Cruikshank's research demonstrates that how one behaves in the environment is heavily influenced by how one thinks about it. In Cruikshank's work, members of the Champagne-Aishik Nation in the southern Yukon have experienced glaciers to be sentient, sensitive, and able to exact revenge for improper observance of protocol and respect in their presence.

Based on her many decades of working with elders and communities in the southern Yukon, Cruikshank argues that "elders talk about the same issues that concern scientists," but they do so with fundamentally different objectives and a sense of causality, often assigning moral failings rather than mechanistic explanations.[9] Cruikshank makes clear that there are ramifications to a differently conceived and symbolized sense of place and landscape. Colonialism allowed for a certain kind of inscription on differentiated landscapes from jungles to glaciers that justified its expansion and tactics toward both landscapes and people.[10] And the divergence of these views remains a point of conflict and misunderstanding in current debates over environmental issues, land claims, and other negotiations or policies that involve indigenous rights (Fienup-Riordan 1990; Nadasdy 2003).

As TK has emerged as a useful node of information, a spectrum of data has emerged as has a spectrum of practitioners, both indigenous and nonindigenous. In climate change–related TK, one is less likely to hear mythology except perhaps as an orienting device. The why things happen question or moral relationship as explanation largely falls out of the equation. Instead, TK is more often deployed as evidence brought alongside scientific facts, but science in the Arctic, as I discovered, is not quite a straightforward endeavor either.

Traditional Knowledge, Social Science, and Science Policy

The inaugural event of the 2007/2008 International Polar Year (most often referred to as IPY) was the Arctic Science Summit Week (ASSW), held at Dartmouth College in Hanover, New Hampshire.[11] ASSW is an annual event, organized and funded by a long list of Arctic-focused organizations. In 2007, that list included the International Arctic Sciences Committee (IASC), Arctic Ocean Sciences Board (AOSB), European Polar Board (EPB), Pacific Arctic Group (PAG), and the Forum of Arctic Research Operators (FARO). Many more science policy and research groups were in attendance as well—all of whom were usually referred to by their acronyms, making it rather difficult to sort out at first for a newcomer to polar science like myself. I often found myself at the back of the lecture halls in which the meetings took place, laptop open, searching on the acronyms in order to try and keep up with the fast pace of abbreviated discussions.

ASSW's program included an in-depth update on the state of Arctic science. Lead researchers working on topics related to permafrost, coastal erosion, ice core data, sea ice measurements, social science (including traditional knowledge), and other areas gave presentations. A fair amount of time was also devoted to presenting and discussing organizational and policy issues. Every polar country was heavily represented as well as some others one might not expect, like Korea and China. The annual website describes the purpose of ASSW thus: "to provide opportunities for international coordination, collaboration, and cooperation in all areas of Arctic science and to combine science and management meetings."

Indigenous presence was few and far between at ASSW. Cochran was there, as was another ICC Canada representative. Yet rhetoric in support of indigenous people in the Arctic could be found in most public presentations and quite explicitly in an exhibit mounted for ASSW at the Hood Museum called *Thin Ice*. The exhibit made available a few of the thousands of items from the Stefansson collection. Vilhjalmur Stefansson was a Canadian-born Arctic explorer and ethnologist and later Dartmouth's director of Polar Studies, as well as a significant contributor/researcher to ASSW's cohost, the U.S. Army Engineer Research and Development Center's Cold Regions Research and Engineering Lab (CRREL), also located in Hanover. The exhibit was curated by Nicole Stuckenberger, a postdoctoral fellow at Dartmouth who had done fieldwork in Qikiqtarjuaq, a small community in Nunavut, Canada. She had gone through the collection and selected a narrative that highlighted changes in the Arctic related to climate change. At ASSW, she gave a tour for scientists attending the conference.

Stuckenberger began by talking about common metaphors we use regarding weather like the cartoon character Snoopy's famous "it was a dark and stormy night" from the *Peanuts* comic series, or 9/11 observations that "the sky was so clear and blue that it seemed nothing bad could happen." Then she explained that the Inuit use myth to understand weather.[12] She used the unfamiliar and almost funny image of a "bad baby" acting up to demonstrate how weather is perceived culturally. Weather, in Inuit cosmology, is like "a bad baby prone to fits," and in times previous, it could be placated by shamans.

In its panels, *Thin Ice* referenced a previously mounted Smithsonian Museum exhibit that had attempted to orient disparate publics by using the metaphor of "a friend."[13]

In recent years, Inuit have described the weather as *uggianqtuq*—a word that suggests unfamiliar, unexpected behavior. The title of a recent exhibition on Arctic climate change at the Smithsonian National Museum of Natural History, *A Friend Acting Strangely*, was inspired by this term. Inuit have described the weather as more unpredictable, storms as more extreme, summer days as hotter, and the land and sea ice as increasingly unfamiliar. Elder Iyerak from the Igloolik Research Center explained the meaning of *uggianqtuq* to an anthropologist as follows: "For example, I am very close with my sister. Say I wasn't feeling myself one day and I went to visit her. As soon as I walk in the room, or say something, she would know right away that something is wrong. . . . She would say that I was not myself."

The metaphor of a friend or the mythological belief of weather as a "bad baby" opens the way for establishing a different relationship with the natural world—one that revolves around hunting primarily, as well as other subsistence activities. *Thin Ice* makes it clear that observation and "knowing when" are the keys to hunting and survival.

Before going hunting, fishing, or berry picking, one has to know where to go and when to start out within the context of the particular season. Such decisions must be based upon traditional knowledge, observations of the weather and snow, wind, and ice conditions, and information from available technologies that measure or help deal with the environment.

It is this hybridity of knowledge and observations that has come to represent TK in the climate change conferences and conversations that I was privy to.

IPY's director, David Carlson, an American, was on my tour of the exhibit. And like everyone on the tour, he paid rapt attention to Stuckenberger as she walked us through the panels. Afterward, he told her that he had recently been to several communities in Arctic Canada, and he was enthused by the eagerness of people to talk about changes they are witnessing there. Stuckenberger agreed, and they traded experiences about elders sharing stories with them. When I later interviewed Carlson, he said that at every meeting he attends, there's always "talk about how it's important to have indigenous partners," but he acknowledged that the rhetoric doesn't always match reality. He said with IPY, they were trying to do better. He noted, in particular, projects by social scientists Igor Krupnik and Sherri Gearhardt-Fox as key exemplars. Krupnik and Fox have initiated long-term collaborations with communities in Alaska and Canada, respectively, in order to systematically record observational data and practices in traditional Inuit communities.

Carlson told me that he addressed relations between TK and science as a problem of valuation and evaluation. He said that what needs to occur for engagement and partnership with indigenous peoples is a rebuilding of the evaluation system so that "what constitutes valuable data" is arrived at through compromise. "Engaging means they [indigenous partners] not only need to be sources of information but they have to set up the structure of what data has value, how we collect it, and how we should share it, and that's a different level of engagement and fairness."

Carlson mentioned Krupnik particularly as a model for thinking about moving beyond indigenous peoples as topics and engaging them as partners. "To understand the Arctic, we have to understand how Arctic people understand the Arctic. But that's not only weather data. That's not only wildlife health data. It's reminiscences, it's language, it's geographical mental maps that are different than geographical physical maps."[14] Here it's clear that Carlson recognizes the epistemological differences as well as the texture, form, and vernaculars that make TK a differently constituted process of knowledge production and expertise.

Partly as a result of Carlson's praise, I sought out Krupnik after his presentation at ASSW and later for an interview in Washington. Krupnik has pioneered both the publishing of this kind of information as well as models for collaboration with communities. Krupnik told me that TK is difficult to get right. It poses significant challenges in terms of its status as a differing system of knowledge, how data in the form of narratives, observations, and beliefs are collected, and how they are integrated with science. He was careful to point out that there are experts acknowledged within most Inuit villages, particularly in Alaska where he's spent much of his research time in recent years. TK is not ubiquitous in Arctic communities. One must have a deep knowledge of the community to be able to ascertain who is an acknowledged and trusted expert.

In his essay in *Watching Weather and Ice Our Way*, a book based on a four-year project on observations regarding climatic changes and weather patterns on and around St. Lawrence Island, Alaska, Krupnik makes the point that his Yupik coauthors, Conrad Oozeva, Chester Noongwook, George Noongwook, and Christina Alowa, are very conversant with scientific terms (2004). They are much more able and eager to integrate scientific findings into their own systematic observations than scientists would be in their encounters with TK. Krupnik argues that the Inuit way of recording, analyzing, and integrating empirical data constitutes a system that can, when done by "experts" among the Inuit, offer long-term localized analysis and prediction, and it remains open to new data like that which science can offer. Expertise is developed over a lifetime and is usually acknowledged and revered by the community or group of communities in which an expert lives. So when hunters need to understand the conditions, they consult their own village or community experts, and in the case of sea ice, they are trusting them with their lives.

Henry Huntington, an Alaska-based anthropologist who has worked extensively with TK, including on ACIA and in other collaborations with Caleb Pungowiyi and Krupnik, said that he has watched attitudes toward TK gradually change. When he first published a peer-reviewed article on TK, it had difficulty passing, but now he regularly reviews articles by others that deal with TK. Huntington pointed me in the direction of a 2000 report based on a conference in Girdwood, Alaska (just outside of Anchorage), held by the Marine Mammal Commission on the Impacts of Changes in Sea Ice and Other Environmental Parameters in the Arctic. Convened by a group of five that included Huntington and Pungowiyi, the goal was to bring together "scientists and indigenous experts" to talk about the changes in the Alaskan Artic. It also included some of the experts and coauthors from Krupnik's "Watching" project. While important changes were documented, commitments to trust, communication, and collaboration also emerged. The report ends on this equalizing note:

> It is almost trivial these days to talk about "barriers" and "hurdles" on the ways Native or local knowledge can be matched with the data collected by the scientific community. Those obstacles most commonly listed arise from the presumption (which more often than not remains untested and never fully examined) that traditional knowledge is assumed to be intuitive, holistic, qualitative, and orally transmitted while academic or scientific knowledge is primarily analytical, compartmentalized, quantitative, and literate (Berkes 1993, Eythorsson 1993, Lalonde 1993, Nadasdy 1999). While there is some truth to these differences, both scientists *and* Native observers can effectively operate with *both types* of knowledge. (45)

Such dual expertise is evident in collaborative projects like those undertaken by Krupnik and Huntington, and it will likely occur more often as figures like Pungowiyi and Krupnik's coauthors become more widely recognized as experts.

Backing up this articulation of an *ideal* of sorts, the continual rhetoric I witnessed in sites like ASSW point to the fact that scientists value the contributions of TK. But the smoothness of pronouncing the existence of such dual expertise and the presence of supportive rhetoric elides the varying processes by which it is occurring. By processes I mean to signal the code-switching, translation, and interpretation that is required for the various mediated and other forums that comprise climate change regimes nationally and globally. In other words, despite the gains made in these specific instances, the status and representation of TK, when considered more broadly in arenas of climate science, reporting, and policy-making, are still very much "up for grabs."

Science magazine's 2007 article on TK features a beautiful image of a Saami reindeer herder in traditional regalia out on the tundra and signals a new and growing acceptance of TK (Couzin 2007). Huntington, Krupnik, and Pungowiyi are featured in the article. It begins with a joke: There are three sure signs of spring. The ducks and the geese coming back, tourists coming back, and scientists who come back to check their instruments.

The joke made me laugh when I first heard it from Pungowiyi during my visit to Kotzebue, but as with all jokes, there's an uncomfortable truth here, too. Scientists are not residents, nor are they invested in community life. Rather, harkening back to Watt-Cloutier's summary of the ACIA process, they are interlopers who may make a contribution to the life of a community, but their goals, norms, and practices differ significantly from those of residents.

Several scientists told me that Nuuk and most of the villages inhabited by Inuit people in Greenland were teasingly called "fly-over" zones, since scientists tended to bypass these major cultural and population centers as they busily headed to their remote research sites further north. Both of these incidental commentaries imply that models for collaboration and partnership have yet to become anything like a widespread norm in Arctic communities.

During the course of my research, I had the opportunity to informally meet Aqqaluk Lynge, 2010–14 international chair of ICC and longtime Greenlandic activist and leader, when he spent time in residence at Dartmouth where he was an invited fellow in 2008. His visit was arranged by

Ross Virginia, a scientist and director of the Institute for Arctic Studies in the Dickey Center for International Understanding at Dartmouth. Virginia was also a lead organizer of ASSW, and I interviewed him at Dartmouth. Virginia specifically said that he invited Lynge in order to move past the "fly-over" problem and engage with communities. But he also said it was something of a risk because ICC "takes strong positions around issues of considering the community. [And] there may be people that disagree with those decisions." But he said, "Having Aqqaluk, I think, enhances the educational experiences for students that are here, and I think it improves the scholarship in some of the programs that we're trying to build by fully understanding and engaging with ICC." He said that as a representative of Dartmouth, he would not "sign on to ICC positions," but as an individual Arctic scientist, he had no problems with their positions.

The Inuit, like other indigenous Arctic peoples, have been forced into engagement with government policies and media—and into forming political representation in order to chart paths of self-determination. ICC was largely formed to engage policy, media, and science—in order to address oil and gas development, whaling quotas, seal hunting and import bans, persistent organic pollutants, and now climate change (Damas 1985; A. Lynge 1993; F. Lynge 1992). Each of these issues is fraught with controversy, conflict, geopolitics, environmental advocacy, engagement with science, and the needs of communities who depend on land-based subsistence activities. Positions, as Virginia put it, are required for a political organization like ICC.

Lynge is not one to shy away from controversy when it comes to challenges confronting the Arctic and Inuit people. In person, he is full of energy, ideas, and passion. When he spoke at the opening of the *Thin Ice* exhibit at Dartmouth, he issued this challenge to those in attendance:

> It is too early to tell how climate change will ultimately affect us. Will the impact of climate change be as powerful and culture-changing as our missionaries and our colonizers were? Will we find the right adaptation measures? I don't know the answer to that. I do know, however, that we will be strong in our resolve to take our own steps in dealing with this. Sometimes we will do it alone, and at other times we will reach out in partnership.
>
> With all the flurry of scientific inquiry on this issue, one could easily be led to believe that it is the researchers who are most affected by the world's changing climate and not the Inuit. I plead with western scientists to be careful how you conduct your research on our land and

on our thinning ice. Work with us as equal partners and not as the colonizers and missionaries did. Help us deal with not only your own interesting research but with our concerns. For example, help us deal with industry, which is keen to see an Arctic sea route open up to them.[15]

For Lynge, then, it's clear that the specter of colonialism not only hovers but can be seen in the traces of how climate change adaptation and mitigation policies are considered and negotiated. It's here that perhaps a rejection of climate change begins to make some sense, as does a continued détente with science. Scientists and their facts are, in this formulation, part of an assemblage with historical antecedents, sediments, and institutions. The hope he offers is in self-determination—that these same figures have the potential now to provide partnership in efforts that constitute successful adaptation to the coming environmental, economic, and social changes. Lynge's presentation echoes both the need for collaboration that IPY's David Carlson earlier alluded to and the push toward self-determination that has marked ICC since its inception.

Carlson agrees that engagement entails not only a commensurability and translation of data but indigenous groups setting the agenda of what gets researched, and he said that at times, it can be frustrating. For example, scientists wanting to study Arctic char have ended up being pulled into politics over housing. Among other scientists I spoke with informally at ASSW, I discovered a range of experiences with communities—some very successful like the Iñupiat community in Barrow, Alaska, that has a decades-long collaborative relationship with U.S. Army and other scientific researchers.[16] A few, however, either expressed skepticism about the rhetoric in support of TK at ASSW or cited research collaborations that either weren't productive or proved difficult to navigate. What IPY offered, however, was one of the largest funding infusions to polar science yet and a coordinated planning effort for polar scientists across the Arctic. Carlson said he wanted to see a different legacy for IPY and more along the lines of what Lynge calls for. He said he didn't know "quite how to do it, but I don't think it's possible to separate science from the policies especially in the North." For Carlson, polar scientific research is grounded in a terrain that not only involves communities but, arguably, *requires* communities. Whether or not such sentiments become the long-term legacy of IPY remains to be seen, even several years out.

"The Right to Be Cold"

Sixteen months prior to the kick-off of IPY, the connection between science, policy, and indigenous communities achieved a major milestone that both built on and exceeded what ACIA had sought to accomplish. On December 7, 2005, at the UNFCCC's eleventh Conference of the Parties (COP 11) in Montreal, a group gathered for a side table session called "The Right to Be Cold." There, Sheila Watt-Cloutier, then international chair of the ICC, articulated what remains a definitive statement regarding how climate change was and is wreaking havoc in Inuit communities across the Arctic.

Beginning in Inuktitut, Watt-Cloutier identified herself by her Inuit name and welcomed the crowd. She then switched to English and acknowledged fellow indigenous people in the audience. Among those who sat at the long head table beside the podium were Inuit hunters, Robert Correll (ACIA chair), James Anaya (an international indigenous law expert and UN special rapporteur on indigenous issues), and Lloyd Axworthy (a former Canadian foreign minister).

After her greetings, Watt-Cloutier announced that, after two years of research, she and sixty-two other Inuit individuals had submitted a petition to the Inter-American Commission on Human Rights. The petition names the United States as a violator of the 1948 Declaration of the Rights and Duties of Man. The petition states that U.S. inaction on reducing greenhouse gas emissions to mitigate the effects of climate change violated the Inuit right to life and physical security, personal property, health, practice of culture, use of land traditionally used and occupied, and the means of subsistence.

The petition was not a surprise to anyone in the room or to those like me who watched via grainy streaming video. In fact, it's arguable that the real weight of the announcement had come the year before in 2004 at COP 10 in Buenos Aires when Andrew Revkin at the *New York Times*, as well as numerous activist and other outlets online, became aware that the petition was being considered and prepared. The headline on Revkin's story, issued December 15, 2004, read, "Eskimos Seek to Recast Global Warming as a Rights Issue." (*Eskimo* is a term used only by Americans. It's considered somewhat derogatory with colonialist overtones in Canada and Greenland.) Revkin interpreted the ICC effort undertaken by Watt-Cloutier as part of a broader turn by "representatives of poor countries and communities—from the Arctic fringes to the atolls of the tropics to the flanks of the Himalayas" who "say they are imperiled by rising temperatures and seas through no fault of their own." Revkin summarized their actions by saying: "They are casting the issue as no longer simply an environmental problem but as an assault on their basic human rights."

The petition serves as a record or oral history of such an "assault." But when I first met Watt-Cloutier in March 2007, she described the petition quite differently—as a "gift." In a speech I attended in Saskatoon, Canada, she elaborated on it this way to the audience of Arctic researchers and academics:

> It [the petition] was not an aggressive act, it was not a confrontational act, and we were actually reaching out, not striking out. It was more of—much more—the powerful assertion of our rights than a lawsuit would have been because I think people would have thought they just want money and that they would have dismissed it. And so we didn't want to go that route. So I always say that *our petition was a gift, a gift from our hunters and our elders to the world.* It is an act of generosity, in fact, from an ancient culture that is deeply, deeply tied to the natural environment and still very much in tune. And it is a gift from us to an urban industrial modern culture that has largely lost its sense of place in position to the natural world. I always say that the petition is the most caring, loving act I have ever brought forward in the protection of my ancient culture, and it is the most loving and caring act I have ever brought forward in the protection of the future of my grandson, who is learning to hunt with his father.

I didn't fully understand the metaphor until I saw about a dozen of the sixty-two petitioners' video depositions included in the *Thin Ice* exhibit at Dartmouth College. In those videos, petitioners describe firsthand what changes they are experiencing, what it means for their families and communities, and the ramifications of these changes for their culture and way of life. It is a staggering testament both to the life of hunting and subsistence living still practiced in the Arctic and to the changes wrought by forces far outside their control.

The video depositions were taken by two undergraduate students (one from Dartmouth), who traveled to remote communities across Canada under the tutelage and with the advance preparation of Watt-Cloutier. She said she would phone ahead and make arrangements, and community leaders would welcome the students, assisting them in setting up and conducting interviews with elders and those considered experts on the topic of climatic changes in their environment. Watt-Cloutier said that fellow petitioners and their communities, primarily in Canada, were unequivocal in their support of the petition.[17]

Regardless of the largesse inherent from the perspective of the petitioners and the landmark nature of the petition, the Inter-American Commission rejected the Inuit petition in late 2006. Watt-Cloutier said that she was devastated when it was rejected. Some I've spoken with have speculated informally about American influence on the commission and the subsequent demise of the claim, but none of these suspicions have or likely could be substantiated. After the rejection, Watt-Cloutier was invited to give a presentation to the commission that would summarize the vulnerabilities globally of indigenous communities to the perils inherent in climate change predictions. This would include the list of "poor countries and communities" Revkin alluded to in his story. Watt-Cloutier told me that a member of the commission had said they wanted to do "something" about this and that her 2007 testimony would help them figure out just what that "something" was. Nothing has come of it since.[18] When I interviewed Revkin, I asked him why he didn't follow up on the rejection of the claims made by the Inuit. He quickly looked in his *Times* database and said that they did follow up—with an eighty-six-word story, noting that the case had been rejected.

Despite the failure of the case and the ways in which it largely fell out of mainstream media coverage, the petition moved the experience of climate change outside of the realm of mere illustration and into the domain of self-determination, power relations, and settled causality. In other words, it isn't greenhouse gas emissions doing this to the Inuit, but the U.S. government, which has stalled on mitigation policies that might prevent further loading of emissions such that climate change will occur more precipitously at the poles, where effects are known to be more extreme. The Inuit are compelled then to deploy a variety of means to address the prospects for their communities' continued survival. In contrast to largely scientific fact–driven appeals in mainstream media, the claim sought to, in Watt-Cloutier's words, "put a human face" on climate change and the Arctic. In so doing, it widened the framework of expertise and of who could speak for and about the issue of climate change in wider public arenas. It made climate change an indigenous, Inuit, and polar issue.

For Watt-Cloutier, this approach began long before she got involved with the issue of climate change. When she became Canada's ICC chair in 1995,[19] studies were coming out that showed how persistent organic pollutants (usually abbreviated to POPs) were circulating from factories in the United States and being found in the Arctic ecosystem, including in the bodies of Inuit people—in fatty tissues and breast milk (Downie and Fenge 2003; Hilts 2000; P. Miller 2000; Sze 2006).[20]

> When I started then, that was the time, very shortly after, when the actual global negotiations were starting on the persistent organic pollutants and the toxins that were coming into our bodies and nursing milk of our mothers and I jumped right in. I hit the ground running with this issue because for me, as a woman, I could certainly relate to nursing milk being poisoned. My daughter was—this was her childbirth age, and so I just felt for the women who would have to think twice about nursing their babies because of poisons coming from afar. Because for us, it was a diet-related issue in the fatty tissues of our marine mammals, and Inuit were most impacted, more than any other aboriginal peoples because we eat seals and whales and walrus, which is where these POPs would make their home.

Watt-Cloutier describes the work on POPs as one of influencing the global community to "do the right thing about toxins," and she explains: "We did it from a high moral ground. We did it from a very human perspective that we were the net recipients of POPs." The UN negotiations on the Stockholm Convention, she noted, were "the fastest UN treaty to have been signed, ratified, and enforced in the history of the UN." In an essay in a book coedited by Terry Fenge (who worked with ICC Canada and Watt-Cloutier on the project), the chair of the UN negotiations noted that he was given an Inuit carving of a mother and child by Watt-Cloutier early on, and he said he kept that concept, embodied in the statue, uppermost in his mind throughout the rest of the negotiations.

Watt-Cloutier began her four-year tenure as the international ICC chair only one year after the Stockholm Convention was signed in 2001, and she said her strategy of "putting a human face" on climate change had its roots directly in the previous seven years she had spent working on POPs. In my interview with her, she explained that she saw climate change and POPs as intertwined because they were both "about health and cultural survival." During her first year as ICC international chair, Watt-Cloutier fund-raised heavily with climate change in mind "because we still have a long way to go in getting the world to understand that this is a human issue." Because of that, one of the avenues she immediately began to explore was the idea of it being a human rights issue.

When she took over the international chair position, she said, the ICC board began asking: "What recourse do we have as Inuit against climate change? How are we protected? We are being poisoned, and now our ice is going to go and our way of life is going to be gone. How are we going to be able to do this?" While these considerations were going on at the board level, Watt-Cloutier considers it serendipitous that in her travels to Washington, D.C., she was able to meet with others who were trying to connect climate change and human rights.

She met first with the Center for International Environment and then Earthjustice (formerly the Sierra Legal Defense Fund). She said she was skeptical at first: "I was thinking, okay, what is this all about? What's in it for them? Are they real? Is it just some new pet project that they want Inuit to get involved in? Do they have potential to really change the discourse on these issues?"

Watt-Cloutier became convinced that a partnership would work. Bringing it before the ICC board, she was met with skepticism and challenges, much like her own initial reaction and particularly about the idea of working with environmental advocates. But eventually the idea of pursuing a human rights case received support.

Throughout the process, she said, there was fear about bringing such a case forward, particularly against the United States. She said many asked her quite pointedly: "What if we wake up the sleeping giant?" Her answer to those questions was equally candid: "That is my point. We are trying to wake up the sleeping giant, and I can guarantee you he's not sleeping. There were a lot of fears involved in moving forward in such a bold and courageous step."

ICC is dependent on government support and funds raised through foundation grants and other donations. So these questions and concerns represented material and structural considerations in taking on this kind of direct confrontation. Challenging the United States on emissions reduction at a time when the Bush administration still claimed that the science was not settled enough to take action certainly would seem to be "waking the sleeping giant." Many environmental advocates cheered on the petition for this reason. It acted in ways that were outside what scientific facts alone could do, by coupling them with facts-on-the-ground in order to convince wide publics. The perceived bias of mainstream media played a central role in how Watt-Cloutier's partners thought about publicizing the petition. When I first contacted an Earthjustice representative about the Inuit case in 2005, he wouldn't speak with me unless I first read the essay "Balance as Bias" by Maxwell Boykoff and Jules Boykoff. The Earthjustice representative wanted me to be aware before speaking with him that the science was indeed settled and had been misrepresented by mainstream media. For Earthjustice, scientific findings were not a sideline but a constituent rationale for Inuit claims, as was the media context in which the petition was re/presented.

In retrospect, Watt-Cloutier describes the two-year period of preparing the petition as a "leadership challenge" where she forged ahead believing in the "honorable intention" of the petition. When it came to signing onto the petition, ICC as an organization opted only to sign a resolution supporting Watt-Cloutier and her sixty-two copetitioners instead of fully joining the petition. Watt-Cloutier pointed out that ICC is a diverse transnational organization that answers in Canada's and Alaska's case to regional development corporations that represent the communities. Many of the corporations in Canada have land claims agreements that involve development, and Alaska is also involved in resource development that includes oil. The concern was that ICC might be considered "hypocrites" if they signed onto the petition. The lack of official ICC backing means that Watt-Cloutier continues to carry on work on Inuit human rights even now that she's out of elected office, and as I noted earlier, she traveled constantly in the years following its submission and rejection to speak about the petition and "the right to be cold."

Politics of Connection

Coupling climate change and POPs together underscores the distinct challenge that Arctic life poses, as residents are both the recipients of industrialization's ills and peripheral players in the policy mechanisms that might stem the tide of such ills. Focusing on the human aspect of chemical compounds and dioxins emitted in the service of industrialized lifestyles is a bold move, but casting it in a human rights framework is much more than a public relations makeover. Michael Ignatieff, a leading human rights scholar and former Canadian politician, has pointed out that human rights are best defended on pragmatic grounds and that there is a fine line between the rights of states and their citizens that must be negotiated in order to protect the legitimacy of the internationalization of human rights norms (2001). So how much an international body could and should intervene in state policies, particularly when that state is the United States and wields an enormous amount of political power and influence, is not a simple proposition for either scholars or pragmatists.

Anthropologist Ronald Niezen has looked specifically at how indigenous political groups have been using human rights standards, particularly in conjunction with United Nations bodies (Maaka and Andersen 2006). He has observed that human rights have become a vehicle for transnational indigenous groups like ICC to pursue self-determination and to enact reform at various levels of law, international organizations, and bureaucracies. He points out that underlying such moves is a tacit agreement that state legal systems cannot be relied on for redress of rights claims. At the same time, however, Niezen points out that human rights frameworks are often unable to cope with claims to difference, group rights, or self-determination due to the antirelativist and individualistic definitions assigned to universalized notions of human rights.

The conundrum for indigenous groups further lies in what some have called "strategic essentialism," where indigenous groups must demonstrate a special relationship with the land in order to have their claims acknowledged (LaDuke 1999). Anthropologist Shepherd Krech has been a vocal critic of these kinds of claims, particularly as they relate to the environment, drawing criticism from many, including indigenous groups, for his lack of acknowledgment and seeming ignorance about the pragmatics of community survival (Krech 2000). Niezen formulates it much differently—as a negotiation between nonindigenous public audiences and indigenous needs that can act as an "artificial boundary."

> Indigenous nationalism thus usually shapes itself around those core values that resonate most strongly with the non-indigenous public. And there is some comfort to be taken in this. Surely there can be little harm in an identity based largely on environmental wisdom. The harm comes more from public disapproval of necessary things, like legal knowledge and resource extraction. An artificial boundary is sometimes erected around indigenous communities that limits their options and inhibits their prosperity. (Maaka and Andersen 2006, 300)

Arctic scholar Carina Keskitalo posed a similar question in her 2004 history of the Arctic Council in which she credits ICC leadership with playing a key role. She asked whether or not special claims to traditional lifestyles foreclosed on the possibilities for a wide range of solutions to deal with social and economic problems. Certainly, this would seem to be a substantive concern in relation to the explanations related to ICC's reluctance to formally back the petition. But as Niezen makes clear, this is not just an Arctic or Inuit concern, and negotiating an indigenous group's public image as well as desires for self-determination, assertions of indigenous title and rights, community survival, and economic aspirations is not a straightforward prospect with right or wrong solutions.[21]

Though she focuses on traditional lifestyles, Watt-Cloutier sees a connection between culture, environment, and community survival. She narrates the lead-up to formulating the petition not as a foreclosure but as a way of opening up possibilities. Climate change projects a wave of devastation for traditional lifestyles in the Arctic and those who rely on subsistence food gathering. Put together with the discovery of POPs, she articulates both a connection with globalized industrial pollution and scientific processes that have increasingly developed the means to substantiate these connections. Her perspective of these global science-driven conclusions as "human," as an *experience* as opposed to a *finding*, and one that is underscored by TK, led her to consider the human rights framework as a means for recourse. Her interest is less in the politics of representations of indigenous peoples and more in a *politics of connection*. She focuses more on what makes humanity similar and approaches solutions from this perspective. Media has played an enormously important role in her approach. She said that as chair, she spent 40 percent of her time fulfilling media requests. She said her approach to media was to say, "You help me tell the story, and I will give you the time to help me tell the story."

When Watt-Cloutier gives a speech, she uses a slide show of often iconic images, some of them awe-inspiring, showing snow, ice, tundra, and Inuit people. Some of the images are recognizable from ACIA, and others are taken by friends or relatives. Many of the subjects are in traditional Inuit dress and depicted outside hunting or traveling across ice and snow. In the course of her speech, she weaves in facts about climate change in the Arctic, painting a picture of rapid transition, globalization, and environmental shifts and dealing as well with the role of ICC, policy, negotiations, and human rights. There is a tacking back and forth like weaving or sailing between introducing largely southern audiences to a "foreign" or exotic world where "ice represents mobility and transportation" and where changes in temperatures can mean a hunter's loss of life or limb. She sometimes describes her neighbor who lost his legs falling through the ice while hunting, or a recent year when the temperature was 8°C when it should have been *minus* 30°C. "The reality is very stark," was how she summed it up in one speech.

The underlying argument Watt-Cloutier is making is that the vast majority of Inuit are exposed to a distinct way of interacting with and understanding the natural environment—because of the very specific environment of the Arctic and its inextricable link to Inuit cultures that have evolved there over millennia. When the majority of the American population lives far south and in urban, industrialized centers, there is a gap to bridge not only between the urban and rural but also between the particularities of the south and life in a far northern climate. Watt-Cloutier builds a case for support of the Inuit, the difficulty of their role in negotiations, the ways in which the earth's environment is connected, and a life lived simply in the cold—"connecting you to the warmth of the ice of the Arctic." Her core message is that "all things are connected," so what happens in the Arctic matters to the rest of the world. She then uses these commonalities to segue back to more familiar territory for the climate-aware, returning to Kyoto negotiations and the world of policy. The meaning of climate change thus shifts toward a form of life that is public and media-savvy enough to present images and stories that evoke empathy, while at the same time reinforcing the factual nature of climate change through on-the-ground examples of how it is already a lived, relevant experience as well as the need for national and transnational political and policy solutions to address it.

What Watt-Cloutier takes away as a definite win is the way in which the petition changed how people think and talk about climate change, human rights, and the Inuit. "It has changed the discourse, there is no doubt about that, and it will continue to do that, but it was not an easy, easy way to go. I wasn't as fearful as some of my colleagues were, thinking something is going to go wrong here and we are going to be stopped and we are going to be laughed at and we are going to be all kinds of things. The reverse happened completely, and that is the trust I had in humanity that the reverse would happen, that people would understand this as a people's right to their way of life that was being jeopardized and it is absolutely."

In contrast to either Krech's or Niezen's observations, then, human rights, in Watt-Cloutier's view, allow for indigenous people to set aside the indigenousness of their claims in order to relate to generic publics as humans, whose lives and livelihoods are threatened. Such a focus on connection doesn't do away with questions about scientific uncertainty, but makes them somewhat irrelevant. It instead evokes the ideal of precaution and communality as well as moral and ethical responsibility. In some ways, she performs a role similar to what Cruikshank describes in relation to TK—Watt-Cloutier assigns moral meanings and not just an explanation of physical mechanisms when it comes to climate change. In so doing, she underscores themes that environmental advocates have been working to advance for decades under the broad rubrics of sustainability. Namely, the petition provides "proof" of industrialization gone terribly wrong, and for those who have already indicted industrialized lifestyles, such a claim provides welcome material proof of the consequences of not heeding earlier warning signals. Yet as Lynge hinted in his *Thin Ice* speech, climate change may indict industrialization, but it might also provide it with its greatest leap forward yet by opening up the Arctic to a level of exploration and development far outside the scope and size previously imagined.

Arctic Rush

The Arctic is certainly no stranger to exploration of either the military or industrial kind, but projections of climate change have catapulted it into a new era of resource potentialities. This doesn't necessarily make Arctic countries more vulnerable; instead, it has the potential to make them and their multinational resource extraction companies much wealthier, which in a group that includes Norway, Canada, and Sweden is hardly a reversal of fortunes. What is more concerning is that these kinds of developments put the indigenous inhabitants in a more precarious position alongside indigenous people of the low-lying nations. How they relate to their nation-states, their distinct cultural ways of being, their relationship to the land, as well as how poised they are to be involved in the political and economic changes predicted in their region of the Arctic become determining factors in their ability to adapt to predicted changes of all kinds.

Robert Correll, chair of ACIA, was at the table with Watt-Cloutier in Montreal when she announced the human rights petition. Correll articulated the stakes of the petition in quite different terms than Watt-Cloutier did. He said, "If you're indigenous people living along the coastal margin, reduction of sea ice is a powerfully difficult thing to absorb. If you're in the oil and gas industry, it opens up pathways that were only dreams some decades ago." Such a formulation makes indigenous people the opposite of rational corporate or state actors bent on massive and steady streams of profit, given that an estimated 25 percent of the world's oil reserves lies beneath the ice (approximately three-quarters in the Russian zone).

Yet as ICC's reluctance to sign on to the petition illustrates, this is not exactly the case, and it harkens back to the warnings offered by Niezen and Keskitalo about what roles are open for indigenous people on the transnational policy stage. In this case, Lynge's *Thin Ice* statement has special relevance on this topic, because how exploration and development play out and what role Inuit people and ICC play in it has, to a great degree, much to do with collective rights, self-determination, and partnerships with science *and* industry.

In the *New York Times* 2005 series "The Big Melt," Watt-Cloutier put it candidly to the series' reporters: "As long as it's ice, nobody cares except us, because we hunt and fish and travel on that ice. However, the minute it starts to thaw and becomes water, then the whole world is interested." The *Times* writers had put it, in contrast, and rather more pointedly in monetary terms: "The Arctic is undergoing nothing less than a great rush for virgin territory and natural resources worth hundreds of billions of dollars." It's worth noting, however, in the face of such fanfare, that as CRREL scientist Jaqueline Richter-Menge noted in several presentations I attended, unprecedented melt (in modern times) of "multi-year ice" (ice that does not melt for five years or more) is not necessarily a linear march to an ice-free Arctic in the summer. In fact, the stable progression of declining multi-year ice cover that would make all of this industrial development possible and guarantee a high return on investment is not something any scientific research can predict. Richter-Menge and James Overland put out an annual report on *The State of the Arctic*, sponsored in part by NOAA, that tracks the relative melt and refreezing, and they are more likely to characterize the future of ice in the Arctic as nonlinear progression where the next ten years may see a major thaw followed by twenty years of renewed multi-year ice cover followed by more thawing.

The view Correll expressed, and what the petition expresses as well, is the dominant mode of representing how these changes will affect Inuit people, but they are certainly not the only view possible. When I traveled in late 2007 to the Arctic Energy Summit (AES) in Anchorage, there were Inuit and other indigenous individuals and delegations there: a permitting group from Barrow, Patricia Cochran from ICC, another woman attached to an environmental advocacy group, and a couple of trained wildlife and fisheries biologists, one of whom had held leadership positions with the Gwiichin Tribe in Alaska. Cochran spoke alongside BP, Shell, and others who were advocating for offshore drilling. She advocated for a view of the Arctic as human as well as resource-based. Iñupiat whalers in Alaska have long been opposed to offshore drilling, so Cochran's place on the program was not exactly a comfortable fit, but it speaks to the way ICC is constantly in a position of negotiating industrial and state forces (that are often mixed in blatant and masked ways).

Alun Anderson, a UK writer I met at AES who was writing about "the Future of the Arctic," later blogged about the ways in which Arctic residents, and especially the Inuit, are depicted as helpless. In a post titled "Get Ready for the Inuit Oil Millionaires," he wrote:

> Right now it is the fashion to see the Inuit people of the Arctic as helpless victims of climate change. It is certainly true that the sea ice is vanishing, weather patterns changing, whales and seals moving to new locations, and traditional hunting lore growing less useful. IPY researchers list many tough challenges. But "victims" they are not. The hunters of the Arctic are about the most resourceful people on Earth. If you can handle a dog team on shifting sea ice in 24-hour winter darkness at temperatures of −40C you know a bit about self-reliance. . . . The story that you don't hear is what the peoples of the Arctic really want: the power to run their own affairs. (Anderson 2008)

He concluded that, although the hurdles are great for self-determination, he wouldn't be surprised if the future of the Arctic included "Inuit oil millionaires alongside resourceful hunters." Anderson is correct in signaling that Inuit people are both resourceful and exploring multiple means for adaptation. Later on that year, as if to affirm Anderson's prediction of resourcefulness, I also met Tony Penikett, one of two negotiators for Nunavut at a conference on the Impact for Diminishing Ice on Naval and Maritime Operations in Washington, D.C. Penikett was the premier of the Yukon territory when indigenous claims were being negotiated there in the late 1980s, and is an expert on Arctic affairs and indigenous rights and claims in Canada. This two-person negotiation team was the lone voice for indigenous people in a room full of naval and policy experts. Their presence and Nunavut's outstanding claims to the seabed acted as a kind of irritation to representatives for Canada who were anxious to shore up their power to negotiate in/for/about the Arctic.

Conceptualizing the Arctic as a region for exploration has a much longer history than these newer stories of what the melting of multi-year ice might portend for industrial development. The Arctic does not fit within the "category" of countries, developing or developed. Instead, following the parameters laid out by the relatively new transnational political organization, the Arctic Council, formed in 1996, spills out over eight nations, 30 million square kilometers, multiple time zones, 4 million people, and thirty indigenous groups. Watt-Cloutier has mentioned this fact on many occasions, but what isn't immediately visible is that the Arctic as region came about as a result of arguments and research done by an international group of scholars and policy-oriented individuals and groups (see Young 1992). Recent historical analysis indicates that ICC also played a pivotal and constant role in the formation of the Arctic Council, and while they failed to get equal billing per country members, they did manage to secure "permanent participant" status for their organization. The Saami, Gwiichin, and other indigenous Arctic groups also participate through this category. The Arctic Council is the political culmination of efforts at region-building and indigenous participation in policymaking, but the identity of the Arctic as ecologically sensitive and distinct was not fully cemented scientifically and within international climate science/policy realms until the release of the ACIA in 2001 (see Martello 2008).

"The Arctic" as entity then remains in the midst of constant negotiation between social, political, and economic forces. It is multifariously defined according to its vulnerability, varying national contexts, economic potential, strategic significance, and mixed populations, as well as its intensive interest to scientists researching climatic change and other issues through a myriad of methods and approaches. These each provide an organizing lens through which the vast expanse of the Arctic can be seen, administered, funded, and coproduced for diverse publics who may or may not pay attention to a polar world considered remote and unknowable until recently.

What ICC does is present a view not from the outside looking in but from and within the Arctic itself. The Arctic as resource looks very different through the prism that ICC representatives present where subsistence hunting and culture revolve around a constancy of ice and snow, self-determination is a constant battle, and traditional knowledge plays a vital role in the understanding of the natural world on a par with science. Getting a seat at the policy, economic, scientific, and international governance tables becomes a crucial part of survival in Arctic politics, and like any political venture, this effort is intricately woven into efforts to capture and mobilize the public imagination as well.

Many of my field sites revealed the continual *crossroads* that ICC leaders are faced with (Fuss 1989). Aspirations for self-determination, economic development, national contexts and histories, international politics, and science have a difficult time staying in their separate bins. At ASSW in particular, I was struck by how scientific findings, challenges, and policy mixed freely with issues of funding, transnationalism, and national retrenchment. TK was something of a darling. Most paid homage to it, and Arctic communities were usually a part of the analysis for the Arctic. There was one glaring and comedic exception of a bureaucrat whose detailed slide show on "regional research policies" neglected any mention of communities. She explained that she had been up late the night before and had forgotten to do "that slide."

But rhetoric belied actual participation from indigenous community leaders. The only indigenous representative on the official program of ASSW (only nonindigenous social scientists presented on TK) was Minnie Grey, a representative of Makivik Corporation, a regional Inuit development corporation in northern Quebec. Flanked by a panel of career scientists and bureaucrats from Canada and the United States, Grey put this challenge to a room packed with international scientists and science policy bureaucrats: "My people have lived for too long with policies that we are not part of. We are slowly being killed by policies that don't help us. Let's create policies together that don't harm our identity." She was the lone voice of passion who personalized the issue of climate change and the driving need to do something—but not just anything about it. The fervently issued plea she closed with was: "Listen to us. Listen to us. We're telling you something is not right."

The human rights petition led by Watt-Cloutier could easily be summarized in ways similar to Grey's message. It is a plea for experiences of climate change already under way to morally and ethically drive public policy, and it acts as a tool for communication, visibility, and connection on behalf of Inuit people. Watt-Cloutier explained to me that she sees public opinion as driving public policy, so her work is continually about tacking back and forth between these worlds. In my terms, then, she is continually pushing the public, media, scientists, and policymakers to expand their notions of climate as a form of life that can include moral and ethical demands, indigenous rights and aspirations for self-determination, potential physical impacts on indigenous ways of life, and the scientific conclusions and predictions that normally define climate change.

Conclusion

I began this chapter by describing the gap between the global fight for attention and the local resistance to "climate change," even in the midst of direct experience with its many symptomatic sets of changes. By seeking to understand that resistance, this chapter has sought to locate Inuit claims within climate change discourse through TK, the human rights petition, and other efforts to address new and transformative development schemes in a warming Arctic. ICC claims on behalf of all Inuit, both formally and rhetorically, espouse two principles: (a) the human in the environment as a constitutive part, and (b) the Arctic as a constitutive part of a global interactive and interdependent ecosystem. Sometimes buried underneath, sometimes front and center, is a parallel principle best described as the right to self-determination—the right of Inuit to have some say in how Inuit affairs are ordered and reordered by trade, pollution, and military/industrial developments in the Arctic and state relations that determine such social, economic, and environmental factors. Inuit claims made through ICC leaders appeal to the universal in order to elevate the particular and are at times both powerless and powerful interlocutors (see Tsing 2005). They are powerless in terms of their non-state status and the remote exoticism often applied to indigenous people and the Arctic, and powerful in terms of the ability to mobilize a transnational network and increasingly, though not without struggle, play pivotal roles in Arctic policy and representations.

ICC brings to the fore the relationship between media, science, politics, and public opinion and, in so doing, performs a multilayered translation. Its spokespeople, like Watt-Cloutier, Cochran, and Lynge, translate the concerns of Inuit communities to the world at large through an array of media and educational outlets as well as the relevance of scientific findings like the IPCC assessments to their own people. Embedded in this process is a push toward self-determination, reclaiming voice, and providing legible representation for a region that has traditionally been defined less by its inhabitants and more by its inhospitable environment, braved by historical expeditions or, more recently, studied by scientist-explorers. ICC leaders perform works of translation and interpretation both to unite an Inuit voice in international and domestic settings and to make that voice heard.

In the next chapter, and throughout the book, the Arctic and indigenous peoples' experience with climate change provide an orienting perspective, heralding what a future with climate change already means. By beginning specifically with Inuit efforts to come to terms with climate change, epistemological differences, and inherently different models for collaboration and intervention, how to both consider *and* live with risk are immediately brought to the fore. Climate change thus becomes both a global and specifically indigenous challenge that is as much a problem of how to define and solve it as it is about how to speak for and about it.

CHAPTER
TWO

Reporting on Climate Change

In June 2007, the University of Oregon (UO) put on
"The Changing Climate Issue: Reporting Ahead of the
Curve," a daylong workshop for reporters. It was titled "Climate Change Boot Camp" when I first heard
about it, and it was sponsored by UO and the Society
for Environmental Journalists. The idea behind it was
that climate change was a story moving from the science pages into all other beats. "More than ever," the
conference description stated, "reporters in every part
of the newsroom must understand some aspect of climate change and explain it to their publics."

Between seventy-five and eighty reporters attended. Most were local reporters in the Pacific Northwest, although a few came from further afield like
Chicago. UO's Bob Doppelt, who is also a local columnist and author on sustainability issues, opened the
conference. He said that the idea for the conference
had come the previous year when he was interviewed

about a report. The resulting story had given equal time to a skeptic. Doppelt called back and asked why. The reporter said they had "googled" and found someone from MIT. The unnamed MIT source called about ten minutes before the story was broadcast, so the reporter decided to quickly conduct an interview and put it in the story.

Climate change as a story, according to scholars, scientists, and journalists, has suffered mightily in the past from these problems of unnecessarily balancing points of view and reporters being dropped into climate change with little or no background on the science and/or debates (Boykoff and Boykoff 2004; DiMento and Doughman 2007; Nisbet and Mooney 2007; Oreskes 2004a; Russell 2008; Ward 2008). The workshop was meant to avert these problems by (1) offering the basics of climate change, which Doppelt described as explaining how scientists and policymakers think and arrive at conclusions, (2) how information tiers down from global conclusions to the Pacific Northwest, and (3) how to accurately cover a fast changing, complex topic.

The program began with renowned climate scientist Stephen Schneider from Stanford University.[1] Schneider began by trying to "distill out the urgency and uncertainty" and said that "what we're really talking about is risk management." He joked that one of the participants was a kid when Schneider first testified before the Congressional Ways and Means Committee in 1976. He said that "back then," it was "all theory." The difference between 1976 and 2007, he said, is that "the last 31 years nature has cooperated with theory . . . the most unequivocal part is that it's warming." He advised journalists to watch out for "myth busters and truth tellers." Instead he summed up scientific results and methods this way: "All good science does not give you answers, it gives you probability distributions," and scientists "worry endlessly about the tails," meaning the extremes or least likely scenarios. He argued that the real debate was not about the science but about fairness and efficiency. He used the melting of the Arctic sea ice as a case in point. It's "terrific" for the shipping industry, which will be able to save on fuel costs by taking more direct and shorter routes across the Arctic, but not so good for the Inuit, who depend on sea ice for their culture. Because the range of global average temperature increase, estimated between one and six degrees, is "not even remotely settled," the questions are really about "how to deploy resources and make decisions with complex science." Schneider was open about the uncertainty that surrounds climate predictions and equally forceful about the need to address those predictions through precautionary decision-making and policies.

During the question-and-answer period following Schneider's presentation, a question came from a radio reporter based in Seattle who said he was not a scientist but an English major who had flunked geology and was having difficulty trying to sort out climate change. He said, "Our job is to give people what they need to find out what's true," and finding out what's true about climate change is no easy task. Schneider responded by advising the reporter that "not all PhDs are created equal" and that skeptics should be given "low status." He said reporters should "do their homework" and learn "whose websites are credible and whose are ideological." It was, he acknowledged, a tough story to cover in a day.

This interaction—scientist explaining area of expertise, journalists seeking how to best cover expert area—is not uncommon for many who work in science journalism. Journalists are expected to learn about the area of research, to converse regularly and develop professional relationships with scientists, to have a sense of how to gauge the impact or newness of the scientific discovery/fact/process, and adjudicate whether and how it merits journalistic coverage.[2] But this interaction and what, arguably, makes it specific to climate change is that (a) nonscience journalists are covering the issue and must familiarize themselves with the issue, its science, experts, and politics, (b) scientific methods, processes, conclusions, and expertise are being vigorously questioned such that some experts are called "ideological," and others deemed "credible," and (c) reporters are compelled to articulate their professional norms as well as rationales for their practices and apply them to climate change specifically. In other words, the credibility of both journalist and scientist are on the line, as are the metrics that measure and account for what and who merits public trust.

Credibility, to a great degree, rests on the norms and practices already built into journalism and science—norms that dictate both what the public good is and how these professions should serve it. Michael Schudson (2001) distinguishes journalism practice from journalistic norms by defining norms as "moral prescriptions for social behavior" and "obligations" or "prescriptive rules" that are "self-consciously articulated." The highly principled moral obligations that journalists dictate for themselves are enshrined in codes of ethics. (The Society of Professional Journalists organizes its code around four key principles: truth-seeking, independence, minimizing harm, and accountability.) As well, most members of the public have expectations and ideals related to what job they think the media should do, and as Jane Singer (2005) has pointed out, many a blog critique reflects such expectations.

Much has been written, too, about scientific norms—first and most famously by Robert Merton in 1942, and particularly since World War II about what the sciences and basic research offer society both as a way of justifying the amount of funding scientific research gets and as a way of explaining and accounting for societal "progress" in the United States (see Bush 1945). Yet even while journalism and scientific norms have been codified and celebrated, they also have morphed over time. Consider the kind of argument that Schneider presents. Climate change–related findings and predictions are not settled or marked by "answers," but they inhabit a spectrum of probabilities—the likelihood of which, he argues, will decrease fairness for some and increase efficiency for others. With such an articulation, Schneider effectively moves climate change into moral and ethical terrain and away from questions about certainty. He trusts the process of science to continue to work at uncovering the riddle that is climate change, but he leaves open whether its predictions can be entrusted to the social, political, and industrial forces that seek to shape the outcomes for their own ends. Science as a self-governing, objective fact-producing set of institutions is maintained, but journalism's role remains in question as it seeks to negotiate social forces and proceed with its watchdog work of holding government and corporations accountable while also educating (and inspiring) the public about climate change and its latent ethical questions.

It's no wonder then that Schneider is met with a question about truth and how to convey what's true. If fairness is what's at stake, then truth must shine the light on the problem, its features, and its predicted impact. The question about how to report truth carries with it an implicit statement about norms: *"our job* is to report the truth"—journalistic norms are thus articulated as a way of explaining the challenge that reporting on climate change presents for journalists who have a responsibility to their profession and, by extension, to democratic publics. Journalists see their primary job as "seeking truth and reporting it"—under which the current Society of Professional Journalists Code of Ethics (adopted in 1996) includes "giving voice to the voiceless." So lack of fairness stemming from an increase in greenhouse gas emissions requires the kind of independent investigation that journalism can *and must* provide.

It's this kind of thinking around and about climate change that has produced an enormous amount of debate about both the role of expertise *and* the role of media in adjudicating that expertise and responsibly informing the public. For journalists, climate change presents a conundrum both in terms of how its attendant facts are represented, stabilized, and mobilized (what "the truth" is) *and* what and how implications and potential impacts should be considered (what "the truth" means). It's in parsing this out that the double bind at the core of this book is most evident—that of needing to both maintain fidelity to scientific expertise and move beyond facts to ask questions about communality, fairness, and what it means to live with knowledge that the future will likely produce more inequality and not less. For social groups, like Inuit leaders in chapter 1 and the evangelicals in the chapter 3, epistemological challenges, vernaculars, and ethical or moral obligations related to a future with climate change are starkly apparent. Yet climate change for journalists necessitates a theory of the social as well and a sense of what it is that journalism should be bringing to the conversation in spite of, *and* because of, its stated professional ethical obligations to uphold democracy by informing the public of "the truth" even "when it's difficult to do so" (SPJ Code of Ethics).

What the workshop conversation and Schneider's advice produce is also a call to report in a specific way such that *the substantiation of facts leads to ethical questions and not more questions about the facts.* And it is on this terrain that negotiating a stance with regard to what I am terming "near-advocacy" is most evident and, I would argue, unavoidable.[3] Advocating for "the truth" related to climate change has been defined as reporting in a way that reifies and relies on scientific consensus and organizes new evidence and findings such that ethical implications emerge. Telling stories such that the ethical becomes the central focus or a central outcome goes beyond "just the facts" and requires an evolution of journalists' relationships with and articulations of traditional norms like balance, objectivity, and accuracy.

Schneider is not alone, nor is it only scientists who have called for different metrics for journalists in the face of both an urgent need to address climate change predictions and organized skepticism that has actively sought to undermine scientific findings backed by widespread consensus. Many scholars have sought to analyze media coverage of climate change in terms of how norms shape what's considered newsworthy and "who speaks for climate."[4] Boykoff's analysis brings to the fore the challenge that journalists face in dealing with climate-related findings, and those who work to sow skepticism such that climate change is called into question as a fact and/or a fact requiring action. This chapter builds on Boykoff and other scholars' work on media by providing what journalism scholar James Carey might deem the other part of the equation when he says, "The appropriate question is not only what kind of world journalism makes, but also what kind of journalists are made in the process" (331). This chapter specifically uses ethnographic data to get inside the issue of how credibility is constructed, perceived, and articulated *by journalists* and particularly how "ethical" journalistic coverage of climate change is debated. It seeks to understand how journalists are being "trained" at workshops and other events that elaborate climate change as a specific science-laden form of life.

This chapter will also address criticism of and by journalists about climate change coverage that can be loosely grouped into a few categories: (1) accusations of bias, alarmist coverage, and exaggeration, (2) claims of inadequate application or explanations of climate science using false balance and ignoring "scientific consensus," and (3) lack of proportionate attention to the issue such that publics might demand and take action. The first, if founded, is a clear violation of journalistic norms around independence, objectivity, and truth-seeking. The second might be seen as ineffective journalistic practices or practices inadequate for the story itself. For example, figuring out how to make the latest IPCC findings "relevant" for a local news outlet is no small feat, though regionalized impact scenarios are growing in numbers. The third, however, presents a distinct challenge to these same norms and practices, and negotiating this challenge is where the gauntlet of near-advocacy becomes clear for professional journalists intent on adhering to norms and to their own sense of what function journalism should responsibly play in society when predictions of the enormity of climate change are on the horizon.

Criticism of journalists and the role of media also gets at a deeper challenge first issued in 1997 by Bill Kovach and Tom Rosenstiel. In their seminal study of public expectations and journalistic norms and practices, they argue that journalists need to be "honest about the nature of what's known and how that knowledge has been generated." Epistemological matters have generally not been a central concern when aiming for a highly professionalized version of "just the facts, please." Yet Kovach and Rosenstiel afford the public both an interest in processes and intersections between knowledge producers and transparency with regard to the choices journalists make in their reporting and analyses—allowing for *an opportunity to hold journalists accountable for which knowledge and which questions they deem salient.*[5]

The problem of how to report on climate change is thus a scientific, moral, epistemic, and existential one—a problem that deals with anticipation, predictions, and conceptions of a future with a range of possible outcomes, produced by a variety of scientific and economic methodologies. So conceived, accusations that journalism is unnecessarily "ringing the alarm bells" for society might be seen as questions about epistemology—about how and which scientific facts are true (to quote the reporter from the workshop) and which knowledges and methodologies matter most in arriving at that truth. Unfortunately then for those who work with deadlines like the reporter who inspired a "climate change boot camp," such forms of life cannot be fully explained merely by trotting out someone considered an expert to speak authoritatively about what is or is not true and/or relevant to the public's interest. You have to, as Schneider put it, "do your homework."

The Reporters' Guide: How to Report on Climate Change

Climate change has produced an enormous amount of "homework" for journalists, policymakers, and the public. In particular, for journalists, the workshop I begin this chapter with is one example, but attempts at cultivating an exceptional set of practices around climate change began much earlier. Following the release of the 2001 IPCC report, which conclusively stated that humans were causing climate change, the Environmental Law Institute released a third edition of *Reporting on Climate Change: Understanding the Science* in 2003. Authored by journalist Bud Ward, now the editor for the *Yale Forum on Climate Change and Media*, it begins this way:

> Like the first two editions that precede it, this is a guide written primarily for journalists. And for other communicators, educators, and just plain "thinkers" who want to take a journalistic approach to the science of global climate change. That is, the kind of approach that adheres to no narrow preconceptions about "who is right?" and "who is wrong?" on the often conflicting science surrounding the "global warming" debate. The kind of approach that recognizes—and respects—the reality that merely striving for "balance" among diametrically competing perspectives may help guarantee just that . . . "balance" . . . but not necessarily the higher standard of accuracy.[6]

Ward signals immediately that there is a "journalistic approach" but that the norm of balance, accuracy, and truth-seeking associated with it shouldn't yield the same kinds of practices as any other story might. Accuracy will be achieved not by setting up the "debate" as a right versus wrong but by understanding the nuance and challenge that a global problem with multiple scientific approaches and key findings presents.

Ward's edited guide leans heavily on the need for scientific knowledge and literacy for journalists and offers ten chapters that summarize the current state of various scientific fields. Ward describes the problem confronting journalists as "an enormous intellectual challenge. It involves all of the 'earth sciences'—physical sciences, life sciences, and some would say even social sciences. It goes way beyond meteorology (the science of weather) and beyond the atmosphere itself."

Reporting on climate change thus requires not only a depth of knowledge on varied fields of research but also an ability to knit differing epistemic approaches together. Rarely, or some would argue never, has an environmental issue enrolled so many disciplines and kinds of research—nor has such an issue been so overtly politicized. Certainly, reporters have not been tasked before with a global science–based issue such that it can and often has subsumed all other environmental issues in a future laced with a wide spectrum of risks. Expertise then presents an experiment for journalists both in terms of navigating their own norms of balance, independence, and accuracy and in terms of translating and representing the science related to the problem. When "the facts" are as complex as those put forward by varied climate models, IPCC and ACIA reports, and other forms of climate change knowledge, independent adjudication and verification of that expertise become a differently configured task, as Ward's opening salvo about "balance" demonstrates.

Ward's articulation foreshadows the oft-cited 2004 Boykoff and Boykoff article, "Balance as Bias," which looked at major newspapers' coverage of climate change and concluded that in an effort to observe professional norms of balancing divergent opinions, reporters had overrepresented skepticism about climate change. In the same year, science historian Naomi Oreskes published her work on scientific consensus in Science and later turned it into an op-ed for the Washington Post. Oreskes found that, in her review of over nine hundred peer-reviewed articles that dealt with climate change, none questioned the basic premise that climate change was occurring. Her evidenced claims to scientific consensus, as well as Boykoff and Boykoff's claim to media bias (because coverage didn't reflect that consensus) produced a kind of unassailable critique of how media had misrepresented the climate story. Oreskes's later work with Eric

Conway in *Merchants of Doubt* (2010) further points out that skeptics have benefited from a strategy titled "teach the controversy," borrowed from evolution/creation debates where less widely accepted and credible views are elevated to equal status under the rubric of teaching all points of view. In this way, doubt gets "produced" via the elevation of experts and the downplaying of widespread consensus. These strategies are particularly difficult to navigate for reporters who are assigned a climate change story without a grounding in its debates and with an overreliance on Internet search engines to find experts—as Bob Doppelt pointed out at the beginning of this chapter in his workshop introduction.

For the reporters at the UO workshop, Schneider described the IPCC as the ultimate navigational tool for following "the signal" and "filtering out the noise" related to climate findings. He said that it is because of the vast amount of evidence collected that the IPCC began its work as a "meta-research council," and its primary task is to weight the literature in order of what evidence is most reliable. The IPCC produced four assessment reports in 1990, 1995, 2001, and 2007; a fifth set of reports has begun to roll out as this book goes to press. The IPCC website has a complex flow chart that shows the process by which it arrives at these reports. The sheer number of authors involved and the long and complex negotiations speak to the difficulty of achieving agreement on what science matters, what that evidence is saying, and what reasonable predictions achieve consensus in order to guide policy.

IPCC puts front and center the matter of expertise and who can speak for and about the signs of climate change. In a funny but poignant moment, Schneider advised reporters at the workshop: "This is not a job for you and your neighbor." In other words, adjudicating scientific research is a job requiring a high level of expertise. Yet, despite the presence and the strength of the IPCC's declarations, particularly from 2001 onward, journalists have continually been asked to do exactly that. It's in this sense that journalists act as a social group vying for the trust of the public along the lines that evangelicals in the next chapter ascribe to "messengers"—those who can be trusted to evaluate the messages of science, scientists, and the discursive conclusions of those vested in certain kinds of policy solutions. But, unlike the coherence that a social group might offer in terms of its translations of the science, epistemological considerations, and use of vernacular, journalists don't have the same sets of resources. Journalists speak for and ascribe to a sense of commonality in epistemology—

in this case, scientific facts and methodologies, and a set of professional norms that are based on an informational theory of democracy—that facts can and should drive action in society.

Andrew Revkin, a veteran journalist turned blogger at the *New York Times*, long associated with his coverage of climate change (he wrote his first book about it in 1992), characterized the problem this way in a 2006 draft of an article he sent me via e-mail:

> Global warming is perhaps the prime example of an environmental is-sue that the media have largely failed to handle in an effective way. . . . *By "effective," I do not just mean accurate.* I mean that we have largely failed to communicate what science can tell us about climate in a man-ner that allows the public to *absorb the information and integrate it* into how decisions are made, both at the personal and societal level. The tendency of the media seems to be either to overplay the sense of im-minent calamity or ignore the issue altogether because it is not black and white. That has left society, like a ship at anchor, swinging cyclically with the tide. And like an anchored ship, we are not going anywhere.

Revkin elevates "effectiveness" to the level of journalistic norm, alongside accuracy. He describes effectiveness as being able to "absorb the infor-mation and integrate it" such that decisions can be made personally and politically. Balance then is not about quelling or creating anxiety *or* doubt about climate change, but rather about the responsibility of reporting such that publics and polities are compelled to become consistently en-gaged and make decisions accordingly.

An evaluation of "effectiveness" is an enduring aspect of critiques and debates about how to report adequately on climate change, and part of it has to do with the ways in which the scientific findings have evolved. Boyce Rensberger, a science reporter with the *New York Times* and then the *Washington Post*, said that when he began reporting on global warm-ing in the early 1990s, the science was a lot more controversial.[7] The Mon-treal Protocol had just come out in 1987 banning chlorofluorocarbons (CFCs), chemicals proven to contribute to the ozone hole. Rensberger had reported on the ozone hole and asked atmospheric scientists working on the ozone whether they thought about the case for global warming. De-spite landmark testimonies in 1986 and 1987, Rensberger said the ozone scientists he spoke with "were fairly skeptical" about the work their col-leagues "down the hall" were doing—the scientists said there were a lot of things they didn't know.

At that time there was a lot of uncertainty and it was completely appropriate for stories to have input that expressed the range of scientific opinion. And so I wrote a story—it was another one of these big package things—that looked at the science behind it. It did not take any alarmist tone or anything like that. What's the evidence, where's the uncertainty, what's the strongest case you can make for it, what's the strongest case you can make against it?

Rensberger said he "got hammered" by "environmental activist groups" for this story—so much so that they called a congressional caucus meeting on global warming that was mostly attended by congressional staff in order to specifically discuss the article. The fear was that the issue wasn't "as cut and dried as they were led to believe," and they wanted to know what the truth was. It was a public meeting, so he went and sat in the back of the room without anyone noticing.

Well-known NASA climate scientist James Hansen was among the speakers at the meeting, and according to Rensberger, Hansen got up and said, "Well, the facts in the article are okay. It's just the *tone.*" Rensberger said he was puzzled by this, and he talked with a lot of other people including Bud Ward and Stephen Schneider. They pointed out that he was focusing on the uncertainties, which Rensberger said "is what a good science reporter does . . . we're trying to give people some basis for judging whether you should believe it [a finding] wholeheartedly, or you should take it with a grain of salt, or whether you should say, 'Well, that's interesting. Let's wait and see how it turns out.'" Rensberger cited Ward, Schneider, and others as pointing out to him that "most environment stories had been written from a sort of whistleblower, alarm calling, watchdog point of view, which is the classic traditional stance of journalists in the United States." Rensberger was quick to point out that this is the reason "why journalism is protected under the constitution. It's supposed to serve the public and be the eyes and ears of the public to report if something is going wrong in the government or anything else that affects us." But, he said, in his stories, "rather than taking that alarmist tone, I just tried to do it straight down the middle." He said that before this experience with global warming, he was even accused of calling the ozone hole a hoax because he said it was a solved problem and "not to worry."

I took a look back at Rensberger's articles for the *Washington Post* and found a 5,311-word story published on the eve of the 1992 Earth Summit in Rio de Janeiro. Rensberger cites the IPCC's 1990 report—the first assessment report as evidence that scientists have not confidently concluded that the rate of warming will be dangerous or that it is caused by humans. He quotes the report as saying, "It is not possible at this time to attribute all, or even a large part, of the observed global-mean warming to the enhanced greenhouse effect [the extra warming attributable to those human-produced gases] on the basis of the observational data currently available." And he points out, "Seldom, in fact, has an issue risen to the top of the international political agenda while the facts of the matter remained so uncertain." He quotes Hansen's 1988 testimony, noting that "the most visible scientists have tended to be those who express alarm and call for immediate, massive action in the name of prudence." S. Fred Singer, a now well known skeptic, was quoted as a severe critic of Hansen who agreed with the IPCC report, calling it "an excellent compilation . . . filled with appropriate cautions and qualifications." With this as a precursor, Rensberger launches into the vast body of the article, establishing it as a guide and "'toolkit' for nonspecialists who believe the future of the planet should be taken seriously." Rensberger walks through many of the details including historical climate shifts, an estimation of emissions, the greenhouse effect, and computer modeling in depth, with a prominence (it ran on page A1) and detail I've rarely seen in a newspaper since I began closely looking at the issue in 2003.

In a history of climate change that includes some analysis and summaries of the media coverage particularly of this period, historian Spencer Weart (2009) notes that most journalists reported on "the issue as if it were a quarrel between two diametrically opposed groups of scientists." Weart argues that this is in part because of efforts made by conservative think tanks, but he also notes that it was "hard to recognize that there was in fact a consensus, shared by most experts—global warming was quite probable although not certain." The latter is definitely where Rensberger said his motivation lay—in the actual lack of consensus on the issue. Weart concludes that "the media got that much right" when they "emphasized the lack of certainty." Indeed, Weart points out, like Schneider did at the workshop, that it was the need for a "better representation of what scientists did and did not understand" that spurred the IPCC to form and continue its work of negotiating and producing consensus statements and views. Yet as the IPCC became more certain about anthropogenic causality and dangerous warming potential with their second report in 1995 and third in 2001, Weart says, media and the public generally paid little attention to the changes. At the same time, industry-funded think tanks and skeptics continued to grow in influence and profile (Hoggan and Littlemore 2009; Oreskes and Conway 2010). It's out of this that the Boykoffs' and Oreskes's research emerged in 2004 and critiques by journalists and authors like Ross Gelbspan and Bill McKibben grew in prominence. Gelbspan went as far as to allege that his journalist colleagues had been duped by or sold out to fossil fuel interests (2004).

This highly charged and critical political atmosphere helps explain the need for a guide like the one Ward wrote and the multiple editions of it (a fourth edition was released in 2012). As Rensberger's story illustrates, reporters needed (and still need) to be able to navigate the scientific research, the institutions publicizing findings, as well as the industry, advocacy, and political interests in order to adequately cover the issue and its ongoing developments—and any fallout that might occur as a result of their reporting. In 2003 Ward began to go one step further than the guide when he worked with Anthony Socci, a scientist with the U.S. Global Change Research Program. Together they embarked on a remarkable series of six two-day invitational workshops for scientists and journalists in order to educate reporters about the state of knowledge on regional and local impacts.[8] Ward told me that with the workshops, they "made a conscious decision to basically fly below the radar stream. . . . We didn't want to seek publicity." Ward posted the links to summaries from the workshops, which are in many ways riveting. They include some of the leading science journalists and scientists (a group that includes Revkin, Rensberger, and Schneider) debating with one another, airing their grievances about Science or Media writ large, and educating each other about their respective professions.

In Ward's book based on the workshops, *Communicating on Climate Change: An Essential Resource for Journalists, Scientists, and Educators*, he says this was the express purpose—for scientists and journalists to educate each other. But he goes further, saying from the start that "frustration was the impetus behind the workshops" (2008, 1). Scientists were roundly frustrated that the media didn't get it and that public engagement suffered as a result. Journalists were similarly discouraged that they still had to convince their editors and the public and battle the rapid pace of change that was transforming their newsrooms, downsizing staff, and putting more demands on their time.[9]

In terms of the work of reporting, balance was an issue that took center stage early on at the workshops. Scientists argued, Ward said, that peer-reviewed articles should not be equally weighted against opinion, policy debates, or political views. At the November 2003 workshop, Ward makes a point of citing Rensberger's affirmation of the growing scientific consensus. Ward writes:

> While there may once have been a legitimate 50/50 split of viewpoints on some climate science questions, Rensberger argued, the preponderance of scientific evidence had since accumulated to a point where responsible reporters should give the scientific consensus on anthropogenic climate change much greater weight than dissenting claims challenging the mainstream scientific conclusions. The journalistic tenet of accuracy now demands that the established science be given total or near total prevalence in coverage of certain aspects of climate change science.

By the time the workshops finished in 2007, this was the dominant view of most journalists I spoke with due in part to a host of likely factors including these workshops, the Boykoffs' article, Al Gore's film, and the fourth IPCC report .

In my interview with Ward, he noted that he thought the workshops "help[ed] create community that certainly journalists knew scientists up-close and personal at a level that they didn't before. They have a much better understanding of each other's issues, including like who writes a headline." In his book, Ward said that scientists were generally surprised to learn that journalists did not write their own headlines—that editors did, and that journalists were quite often frustrated with this process and its outcome. He said it was a bonding moment as scientists also bemoaned the way their universities' public relations staff oversold and sometimes mischaracterized their research with press releases. This community and the trust-building process are not an insignificant by-product, and in many ways they spawned other efforts like the UO workshop I began this chapter with.

Telling the Story:
Journalistic Practice Meets Hurricanes and the Arctic

Climate change, as this recent history of guides and workshops illustrates, presents a genuine and evolving challenge as a news story—what scholars might differentiate as journalism practices. On one hand are the ethical and near-advocacy related challenges: how to present a long-term uncertain issue like climate change that requires action and engagement without sacrificing journalistic norms of objectivity and non-advocacy. Skeptics provide a kind of specter or counterpoint to many of the actions taken by those, especially journalists, who seek to present climate change as a fact requiring action. Part of the complaint of some prominent skeptics has to do with how facts are evidenced. Skeptics tend to favor empirical meteorological modes of compiling and projecting data, while climate science more often relies on more complex models and simulations that enroll empirical and theoretical data to arrive at a range of predictions (Edwards 1999; Lahsen 2008; Mooney 2007; van der Sluijs et al. 1998). Finding one's way around climate science then is part of the ethical and substantive task that is part of a climate story.

But the other set of challenges have to do with the mechanics related to forms and styles of journalism (Broersma 2010). Many journalists, including those at the workshops, note that climate change is a story that "oozes" and doesn't "break." In other words, it doesn't quite fit the mold of what is characterized as news primarily because it isn't happening on a timescale or in ways that demand immediate attention. And finding a picture that illustrates conclusive proof of the fact that climate change has begun is nearly impossible, although the maps showing the decline of the Greenland ice sheet come close. Climate change also defies the framework most have developed for thinking about weather as an empirical, felt experience. It relies on statistics, theory, a wide range of evidence and research, and global modeling to make a case for massive disruptive changes that will introduce a range of variabilities that may or may not begin happening immediately. With the exception of most glacial melt and sea level rise, it may be difficult to recognize them, in most cases, as conclusively connected to the notion of climate change.

Yet the norms of storytelling for news require that journalists find a way to make an esoteric, futuristic concept like climate change relevant, concrete, visible, and legible for the average reader/viewer/listener. Such journalistic dictates stem in part from the democratic ideal of an informed citizenry being given the opportunity through media coverage, as Revkin put it earlier, "to integrate" information into their lives. With media changes and its forms in flux, ever pressed for space and time for analysis, complex issues like climate change present some distinct challenges and opportunities when events that qualify as (breaking) news present themselves.

Hurricane Katrina provides a case in point. Katrina was a larger, more catastrophic hurricane than had previously been witnessed in the Gulf of Mexico, cutting a wide swath of tragedy throughout the Gulf and destroying much of the city of New Orleans. Shortly before it hit, MIT atmospheric scientist and leading hurricane expert Kerry Emanuel had published an article in *Nature* saying that it was likely, based on his modeling, that climate change would increase the intensity (not frequency) of hurricanes. In the days following Katrina, he said his phone rang continuously with journalists looking to make the connection between climate change and hurricanes. *Time* magazine's first cover in the aftermath read, "Are we making hurricanes worse? The impact of global warming. The cost of coastal development." *Time* wasn't alone; many news outlets ran with the story, some even making a distinction between intensity and frequency. Ross Gelbspan published an op-ed in the *Boston Globe* entitled "Katrina's real name." Al Gore's film built heavily on the devastating images wrought by Katrina's destruction. It would seem that Katrina was the first catastrophe that could be considered evidence of climate change, a portent of future risk, and a reason to act now. This was certainly evident in my research with Ceres, as I detail in chapter 5, and in the use of weather-related destruction costs by insurance industry reports in 2005—the year Katrina hit.

Neither Emanuel nor any of his scientific colleagues would say Katrina's ferocity was a direct product of climate change.[10] The Gulf waters were warmer, which likely increased Katrina's intensity. But that wasn't necessarily *caused* by climate change. Indeed, what Emanuel points to as a problem for all hurricane-prone areas is inappropriate coastal development. And what later was revealed to be a primary issue in the destruction of New Orleans was the state of the levees (McQuaid and Schleifstein 2006). Yet it's still possible to point to Katrina as an example of what the globe could be in for in the future.

A number of scientists have turned to blogging, particularly those involved in climate research. Noted climate scientists Stefan Rahmstorf, Michael Mann, Rasmus Benestad, Gavin Schmidt, and William Connolley coauthored the following explanation on their blog, *RealClimate* (subtitled *climate science from climate scientists*):

> Due to this semi-random nature of weather, it is wrong to blame any one event such as Katrina specifically on global warming—and of course it is just as indefensible to blame Katrina on a long-term natural cycle in the climate. Yet this is not the right way to frame the question. As we have also pointed out in previous posts, we *can* indeed draw some important conclusions about the links between hurricane activity and global warming in a *statistical* sense. The situation is anal-

ogous to rolling loaded dice: one could, if one was so inclined, construct a set of dice where sixes occur twice as often as normal. But if you were to roll a six using these dice, you could not blame it specifically on the fact that the dice had been loaded. Half of the sixes would have occurred anyway, even with normal dice. Loading the dice simply doubled the odds. In the same manner, while we cannot draw firm conclusions about one single hurricane, we *can* draw some conclusions about hurricanes more generally. In particular, the available scientific evidence indicates that it is likely that global warming will make—and possibly already is making—those hurricanes that form more destructive than they otherwise would have been.

Rahmstorf et al. separate Katrina from climate change in terms of causal effect, but they don't let it go as an object lesson. Rather, they employ a different frame or set of questions that enroll Katrina as an example rather than an effect. It's a subtle change, but one that still allows for the ethical discussions about what climate change portends in the wake of Katrina. Much like Schneider's earlier characterization, Rahmstorf et al. seek to explain climate in terms of probability distributions and the ongoing processes of scientific research as just that: ongoing. Then they close not with answers but with ethical questions that research brings to the fore for them as scientists—questions that point to eventual winners and losers, where those in hurricane zones will likely suffer as a result of more destructive results. They close with this sentence: "What we need to discuss is not what caused Katrina, but the likelihood that global warming will make hurricanes even worse in future." In other words, Katrina presents itself as a harbinger of an anticipated future.

In the year following Katrina, another research team weighed in—but not just on the legitimate scientific disagreement that Rahmstorf et al. only hint at in their post. Judith Curry, Greg Holland, and Peter Webster published a paper in 2006 in the *American Meteorological Society* that sought to both characterize scientific findings related to hurricanes and climate change *and* their experience with media who sought them out as experts. Like Emanuel, Curry et al. had also published a paper in advance of Katrina (Webster et al. in *Science*) that had their phones ringing constantly with journalists looking to clarify the link between hurricanes and climate change. Recognizing that major media interest would be related to Katrina, Curry et al. drafted a press release that took this into account, and they excerpted this key portion in the AMS article:

The key inference from our study of relevance here is that storms like Katrina should not be regarded as a "once-in-a-lifetime" event in the coming decades, but may become more frequent. This suggests that risk assessment is needed for all coastal cities in the southern and southeastern U.S. . . . The southeastern U.S. needs to begin planning to manage the increased risk of category-5 hurricanes.

Much like Rahmstorf et al.'s blog post, Curry et al. wanted to have a conversation about what to do about storms, and they expected that peer-reviewed research like their article would be the "gold standard" of evaluation. Instead, their interview clips and excerpts were placed alongside climate skeptics of many kinds, and much of the reporting focused on whether or not climate change was real and could be the cause of Katrina. More devastating, however, was the division that media coverage caused within the scientific community. They refer, without details, to misrepresentation of disagreements and unsubstantiated feuds between scientists that have disrupted normally "collegial" relationships. Curry, a participant in the workshops led by Ward and Socci, cites both Boykoff and Boykoff's "Balance as Bias" article as well as the workshop reports as a way to understand how media work and what media are doing wrong. But she and her coworkers also cite fundamental fault lines in the norms, expectations, and epistemic goals of scientists and journalists.

While responsible journalists and respected scientists share some similarities in their "pursuit of truth," they have different and sometimes incompatible goals, missions, and responsibilities. Journalists are not simply looking for information; they are looking to develop stories that are timely and relevant, are wide in scope, have a particular thematic angle, reflect conflict, and demonstrate human drama.

Curry et al. reserve some distinctions for those explicitly committed to science journalism as opposed to political journalism, but in general their sense of unfair, inaccurate coverage is palpable. Much like Boykoff's later work in *Who Speaks for the Climate*, they view journalists as truth-seekers, but with a storytelling mission rather than one of information transfer. As Rahmstorf et al.'s blog post also illustrates, connecting hurricanes to climate change does present a serious conundrum for journalists in terms of providing the public with clear predictions that warrant action, framing relevant questions, characterizing and including relevant scientific findings, nuanced summarizing of probabilities, and connecting global climate research models to single weather events. And certainly Curry et al. as well as Emanuel bore some of the brunt of this challenge in the wake of Katrina.

In contrast to hurricanes, the Arctic provides the most immediate, reliable evidence of current climatic changes and their effects. Drastic images of the melting polar ice cap make for dramatic evidence of climate change. The image of Greenland's receding ice cover year after year shows a clear and present trend toward warming; it was a much circulated image from the 2001 IPCC report. Charismatic megafauna, like polar bears, also play a lead role in stories about the Arctic. *Time* magazine's iconic cover in 2006 was titled "Be Worried. Be Very Worried." Classically written as a hook or peg that makes climate change present for the reader, the subheadline underneath reads: "Climate change isn't some vague future problem—it's already damaging the planet at an alarming pace. Here's how it affects you, your kids, and their kids as well." Beside it is a lone polar bear stranded on an ice floe in the middle of melting waters. Polar bears were placed on the endangered species list as a "threatened species" in 2008 as a result of climate change predictions. Al Gore's film has a dramatic animation of a polar bear drowning because it has run out of energy trying to find another ice floe to rest on. A less popularized aspect of this issue is that Inuit people, particularly in Canada, were upset by the listing, as polar bears are not yet *actually* endangered in terms of current statistical measures, and they are the basis for robust hunting and guiding businesses throughout the Canadian part of the Arctic (Palin 2008; Watt-Cloutier 2007).

Revkin has reported extensively on the Arctic, traveling there with scientists to cover climate change research. He wrote a book about the North Pole for kids in 2006. And as I detail in the previous chapter, he also broke the news of the Inuit claim in an article he wrote in 2005. In a 2006 interview he gave to Brooke Gladstone for NPR's *On the Media*, Revkin had this to say about the Arctic and corresponding sea level rise:

> When you look ahead at the Arctic later this century, there's not a scientist around studying this stuff who doesn't see the prospect of basically a blue pole at the top of the world for the first time in human history, meaning summertime open water ocean, just like the Atlantic or the Pacific, all the ice gone. But when you look at the near term, there's been a lot of melting, a lot of strange things going on with the sea ice that they can't ascribe this particular year to our influence on the climate system. They know it's contributing to change, but there's enough variability in the Arctic that you can't make a slam dunk case.

So that's a nightmare for the media. You know, my editors—the one thing that makes them glaze over immediately is the word "incremental." That's like, at the *Times*, and I'm sure any other newsroom, that's a death sentence for a story.

In other words, Revkin sees it quite differently than those who wrote the headlines for *Time* magazine and tagged it to a polar bear on the precipice of possibly drowning. Moreover, while incremental is what Revkin says is the primary concern, variability could well be a bigger death knell for climate coverage.

In a session I sat in at a conference titled "The Impact of Diminishing Ice on Maritime and Naval Operations" in Washington, D.C., as well as at the Arctic Science Summit Week, I heard U.S. Army Cold Regions Research and Engineering Laboratory (CRREL) scientist Jacqueline Richter-Menge speak about the state of Arctic sea ice cover. Richter-Menge and Jim Overland from NOAA's Pacific Marine Environmental Laboratory are lead authors of the 2006 *State of the Arctic* report and subsequent "report cards." Their presentation includes a dramatic time-lapse animation of sea ice recession. Whenever I have seen their sea ice melt presentations, the evidence seems overwhelming, even to my seasoned eye. Multi-year ice, the thickest ice cover, has been melting at a previously unfathomable rate. The Arctic waters freeze up in the winter again, but that ice is not as strong or thick as multi-year ice. Richter-Menge is careful to say that what she was presenting is what they are witnessing now. She says very clearly that they don't know what the future holds, and that it is possible that the sea ice could freeze up again and stay frozen for ten or fifteen years and then melt off like this again. Speaking as she was in Washington, D.C., to people focused on infrastructure in the Arctic, both industrial and military, this is not exactly the kind of stable news one might hope for if, for instance, one were looking to support either a new shipping route through the Northwest Passage or new polar tourism like the cruises around Greenland.

So even in the Arctic where evidence is definitely associated with climatic changes, the variability throws journalists a curve ball. Change is definitely occurring, but what that looks and feels like for global and regional infrastructural needs, not to mention geopolitical games (of which there are many in the Arctic), remains indeterminate.[11] If we think in terms of journalism practice only, one of the first things students are taught is "news judgment"—what makes a news story. Drama, personalization, and novelty, as Boykoff and Curry et al. both point out, are part of it, but a story won't fly unless there's a news rationale *and* clear evidence to support assertions and claims. What constitutes evidence is where epistemological questions begin and where, as Jasanoff argues, distinct cultural factors also act to determine what matters and who counts as an expert. Yet news rationale is generally explained in terms of *impact and timeliness*—how much it affects wide swaths of the public or "the public interest" and when. Reporters are trained (and socialized) to identify and articulate these elements in order to get approval from editors to pursue a story, and headlines (usually devised by editors and not necessarily the assigning editor, who would have approved the story) also use this rubric in order to sum up impact and timeliness in a pithy short phrase that makes readers or viewers pay attention, click, and/or read and watch the story. But if we think in terms of norms: how much or rather *should* journalism in its myriad of platform-based variations be used to educate the public? Whether the form and institutional structure allow for education, robust participation that questions, verifies, and debates journalistic assertions and reports and/or alternative experts and knowledge are all evolving, open questions.

Form and Structure: Educating, Informing, and Participation

Science, in general, is most often reported at the national level, and even then, national reporting is caught in the midst of industrywide flux. Immediately following the period of my fieldwork, in 2009, the death of newspapers was feared to such a great degree that Senate committee hearings were held to discuss their demise. No such hearing was precipitated when science sections were cut out of newspapers. The *Boston Globe*, for example, decided to cancel its science section well before the hearings (Russell 2009a). Its reporters remained on staff. Despite the rising proliferation and complexity of science issues, the *New York Times*, the *Washington Post*, and wire services are among the few that maintain reporting staff with a science beat.

So part of the challenge confronting science journalism in general and climate change coverage in particular is structural, and part of it is practice-related as the previous section illustrates. Even with all of the guides and workshops and attempts at changing the way climate is reported on, many journalists are focused on audience-specific events and concerns, or their publications don't provide space for reporting on science. But a more basic question remains alongside and undergirds the structural and practice-related ones, and that is the question of what role media should play (and how). For many climate advocates, and for scientists like Curry, Webster, and Holland, journalists can and should be educating the public about the complexity of climate change, the adjudication of expertise, and the processes and methods by which scientists arrive at findings. But when journalists do choose to take such an "educational" stance, it can be perceived as advocacy by their audiences or, worse, propaganda.

The last panel of the UO workshop for journalists was set up to dissect the fallout from a small chain of newspapers that had done a climate series and won awards for their coverage of environmental issues. Their readers, mostly an audience of farmers, said little until the chain publicized their awards, and then the phones began to ring. The editor on the panel described a man who spent an hour on the phone with every member of management staff and the reporters involved, airing his dislike for the series. A skeptic, yes, but it was also an argument about relevance, about how such a big issue fit within the vernacular of farming and everyday life in rural Oregon. It is an argument similar to the one made in Kotzebue that I examined in chapter 1: namely, that even when climate change symptoms are obvious and felt, assigning them to this thing called "climate change" requires translation from one or many vernaculars to others. For reporters, this process lays down a challenging gauntlet requiring various kinds of negotiations inside and outside the newsroom.

Unfortunately for the local Oregon paper, the farmer who vociferously complained wasn't alone. Some subscriptions were not renewed or were canceled outright, and the editor wondered out loud how long this trend would continue, whether it was short term, and whether all the nonrenewals and cancelations were related to the series. There weren't any suspicions that it was an orchestrated campaign—rather, that skepticism had both trickled up and trickled down. In other words, farmers had formed opinions through unspecified means (social, media, social media, or otherwise) unrelated to the local paper so that when the paper presented its take on climate change, it was met with anger and disagreement.

Often local reporters encounter similar resistance long before it gets to the public—from their editors. At the 2007 Society for Environmental Journalists (SEJ) meeting at Stanford University, one of the most striking panels I attended had nothing to do with climate. It was a panel of reporters from places like Tallahassee, Bar Harbor, and Colorado Springs who were speaking about reporting on the environment in a conservative media market. One reporter told of reporting on the governor of Maine attending a special screening of *An Inconvenient Truth*, a newsworthy event in terms of policy and lawmaker influence. Afterward, however, an e-mail went out to all staff from the editor saying that he didn't want any more reporting on climate change "until Bar Harbor is under water."

The sentiments expressed by Oregonian farmers and a Maine editor reveal that for local media, the stakes are much higher than they are for a national media outlet. When naysayers and skeptics weigh in, it's not participation that gets recorded at the local paper. It's cancellation. In other words, at this level, it's not just a matter of navigating expertise and varied scientific research and predictions, although that is an issue as the UO workshop and recent investigations into the strategy of skeptics can attest (Hoggan and Littlemore 2009).[12] Rather, climate change's form of life and how it is understood (or not) as meaningful and relevant play a far greater role for their audiences. Science is easily ignored or shut out by those uninterested in the stakes being pursued by either scientists or policymakers, and that is reflected in the structure of how science gets reported in the American news industry.

When I went to speak with James McCarthy, a well-known scientist at Harvard who chaired the IPCC Working Group Two for the 2001 assessment report and is president of the AAAS, he encouraged me to talk with Cornelia Dean, a science writer for the *New York Times* who teaches science students at Harvard about media.[13] McCarthy and Dean had offices across the hall from each other at Harvard at the time of my interviews. McCarthy said that Dean and he had many conversations/debates about the duty of journalists with regard to educating and informing the public. McCarthy, like many scientists I talked to, saw journalists as educators, but he said Dean drew a fine distinction between educating and informing the public.

When I asked Dean about this, she said: "I think the responsibility of the journalist is to give the news, and what I've said is if people end up learning something in the process, I do not object to that, but my job is to give the news." She said people have to be able to discern right up front, in the first three paragraphs, why a story is important.

> It's very easy in science journalism to lapse into writing what is going to sound like an encyclopedia entry. And it is my belief, untested, that people are not necessarily going to be engaged by encyclopedia entries the way they will be engaged by news. Now very often you'll write a

story about something and there will be a little sidebar that says, you know, the chemistry of the atmosphere, or the life cycle of the whatever, or a graphic that explains it. And so you're educating people, but what we're actually doing, I would say, is giving them the background they need to understand the news that we are telling them about.

As Dean articulates it, then, the duty of journalism is caught up in the norms and expectations of journalistic storytelling and the conventions of what is perceived as news.[14] She said that science reporters are particularly challenged by this because they "have to assume much more ignorance" than on other news beats. She used the example of DNA to illustrate the concept of a "headline word." This is something that ceases to need an explanation. Twenty years ago, she said, DNA needed to be explained much like RNA would now, but somewhere along the way, it became a part of what was assumed knowledge—it became a headline word. She said a colleague at the *New York Times* uses a sports metaphor to explain what science reporters are up against: "It would be as if you were writing about sports and every time you wrote a baseball story you had to tell your readers what first base is." So education is part of a reporter's task, but only insofar as it furthers the goal of explaining what is news about a particular area of research.

This notion of education is connected to a supposed dearth of science literacy among the public. I have heard repeated calls to do better with American schooling in order to get Americans properly engaged with science at a young age. Some journalists I spoke with say they aim for a grade 7, 9, or 12 level of science education. Others, like Dean, say they are writing for a reasonably curious adult reader. Journalist and author Chris Mooney and social scientist Matthew Nisbet have made the point that scientists often think that educating the public (or journalists) means making them think or see what scientists do. Then the public might come to the same conclusions, and controversies would evaporate (2007). But what this chapter illustrates is that, even on a challenging subject like climate change, this is decidedly not the role journalists covering science see for themselves, as either parrots or cheerleaders.

In contrast to Dorothy Nelkin's findings in the early 1990s, journalists who cover science regularly for leading publications are no longer likely to cover science with unbridled enthusiasm. Boyce Rensberger, who began as a science reporter in the 1960s, said there's been an enormous change in the professionalization of science reporting and with science itself.

In the 1960s, science reporters were largely people who saw themselves as translating what scientists do for the general public. They took in the science in one ear or maybe in both ears and then processed it and typed it out in some simpler form. I had an editor once who talked about *running it through the simple machine.* So from that it would come out in a form that—this was in Detroit where there is a big Polish population—so it was Mrs. Poppazuski was the one who had to understand what we were writing about: a Polish immigrant who was more concerned with day-to-day survival than bigger issues. And so science reporters and medical reporters took press releases and announcements from scientists, looked at the journal sometimes, and just wrote—took it all at face value. Today, that's very different. The journals are covered much more closely. Science reporters are much more knowledgeable about science. They are much more skeptical about it. They know that scientists make all kinds of claims, some of which are responsible but highly uncertain by definition. Cutting-edge science is looking into things that we don't know much about; therefore, it's highly uncertain.

Certainly American views of science coming out of the postwar period were focused on the progress that technology offered, and there was an eagerness for news of "discoveries" or, as Rensberger put it, "amazing breakthroughs." Breakthroughs are still sought after and amply reported on. Indeed, the hype is what is often required to fund research that might alleviate critical medical problems (Burri and Dumit 2007; Sunder-Rajan 2006). But that trope has been affected by issues like the threat of "nuclear winter," which turned out to be more of an "autumn," and medical advice that turned out to be incorrect or damaging for the purpose it was intended, as in the case of thalidomide.

Climate skepticism builds on the erosion of the authority of science, which as I detail in the next chapter on Creation Care is definitely greater among some social groups than others owing to historical relations with scientists and core concepts and principles like evolution. But for those without overarching religious beliefs, trust in science is not automatic either. Part of the problem is that science as a process—often two steps forward, and one step back—is not the usual purview. And as Revkin points out, "For every PhD, there is an equal and opposite PhD," and adjudicating expertise has become that much more complicated as claims and counterclaims need to be sorted more carefully if accuracy is the goal and not mere balance.

But it's also about what Dean so articulately points out in Kuhnian fashion: "All science is provisional. It's capable of being overturned." She explained science as a process: "Science looks in nature to answer questions about nature and test those answers with observation and experimentation." Science as provisional means that there will be moments where science errs, and figuring out those moments is a challenge both for scientists and for those who report on them. That science is a process dependent on errors and failures and not a search for solutions is still subterranean even in the formulation of it as provisional.

In recognizing this, and in building relationships with scientists, journalists who engage with climate change are faced with the challenge of navigating balance, independence, and objectivity in their pursuance of truth. Climate change reporting has often been accused of advocacy, particularly by those who have a vested interest in making sure it remains off the radar of the American public. Such assaults fly in the face of professional norms and the journalistic tenet of independence and the trust journalists work to build with their audiences. But they also work to open up these norms for historical and current scrutiny.

Sorting Norms: Objectivity, Advocacy, Truth-Seeking

For those who report on the environment beat, there is a spectrum of beliefs on how to navigate their own role in relation to advocacy.[15] Ross Gelbspan has moved over, according to some, to being an advocate, and so has Elizabeth Kolbert, whose New Yorker series was turned into the book Field Notes from a Catastrophe. Chris Mooney, author of Storm World and The Republican War on Science, told me that he sees part of his work as being education and another part as advocacy in addition to journalism. But it was Dean who articulated the position of those avowedly against any connection with advocacy. When she wrote Against the Tide: The Battle for America's Beaches in 2001, she said she was very careful about being perceived as an advocate on the highly contentious issues she covers in this book.

> I wanted to write a book that would present information that I thought people ought to know about when they consider what they should do on the coast. I have a personal opinion but there are very few things that people have no opinions about, right? I wanted to make it impossible for people with another opinion to dismiss my book as the work of an advocate. You have to inoculate yourself against the possibility that

someone is going to say there's no reason to pay attention to that—we know where she stands. I think the journalist in some ways has the same problem as the scientist. If you become known as an advocate, people will tend to dismiss what you're saying as having been precooked.

Dean, in a follow-up conversation by e-mail, said that this did not mean she gave "equal weight to all sides." Her book came to be seen as an account of the "negative consequences" that have resulted from "many of our coastal development practices."

Advocacy is intimately connected then to criticism about balance and to the long-held norm of objectivity. In Dean's formulation, being seen as an advocate reduces a journalist's ability to adjudicate expertise and the impact their work might have on the widest possible audience. Yet facts like those associated with climate change are compelling as are the ethical dimensions and risks associated with these facts. Good practices associated with balance mean representing a consensus view of science, but as this chapter illustrates, other questions continue to proliferate around what tone to use, how to tell the story, how to hold science experts to account, and whether or not to force ethical questions into the foreground when scientists say the facts demand such questions. In sorting through these problems, a spectrum of near-advocacy has developed as Dean, Mooney, Gelbspan, and Kolbert aptly represent. But in the critique around why the media has not managed to inspire the public to care about climate change, there is an underlying misunderstanding, or perhaps an emergent debate about what it is that journalism can and cannot do according to its own professional standards, norms, and practices. This is what forces many journalists to engage in an articulation of norms in order to demarcate what it is that journalists can and cannot do in response to climate change and to define what kind of challenge climate change's in-flux form of life presents to journalists.

In Schudson's excavation of the objectivity norm in American journalism (2001), drawing on theories from Émile Durkheim and Max Weber, he considers four conditions that encourage the articulation of norms: (1) during forms or events related to "ritual solidarity" for the group, (2) during "cultural contact and conflict," (3) in large institutional settings such that informal socialization is not enough and prescribed rules must be formally generated and circulated, and (4) when superiors in large institutional cultures need to control subordinates in a complex organization. The first two relate to Durkheim's notion of social cohesion, and the latter two relate to Weber's ideas about social control. Schudson identifies the presence of these conditions in the late nineteenth century as the objectivity norm began to take hold. He argues that objectivity was "already operating in the daily activities of American journalists" before it was enshrined in codes and articulated in the 1920s and 1930s and that its emergence is linked to two overlapping impulses within the profession. Journalists at that time "sought to affiliate with the prestige of science, efficiency, and Progressive reform" and "sought to disaffiliate from the public relations specialists and propagandists who were suddenly all around them" (1998, 162). It's into these conditions, Schudson points out, that Joseph Pulitzer and Walter Lippmann begin to advocate for professionalism, scrupulous methods, and scientific ideals. Objectivity as a specifically American journalistic norm becomes a way of defending and guiding journalism even while many recognize its limitations and regard the emergent "interpretive journalism" as necessary in an "increasingly complex" society. Interpretive journalism still required the professional distance and methodology to assess and opine on news of the day, but it allowed for latitude beyond "just the facts."

Objectivity was removed, with much fanfare, from the Society of Professional Journalists' Code of Ethics in 1996. Yet as Stephen J. A. Ward (2004) argues in his history of objectivity, a commitment to "objective methods" persists, and objective is often what is meant when terms like "fair," "accurate," "independent," and "unbiased" are deployed to describe what sets professional journalism apart from bloggers or others who espouse a kind of "affirmation journalism." Affirmation journalism seems like an oxymoron, but it's been adopted to describe those who explicitly interpret news and events through a lens that's roundly seen as ideological. Journalists of all stripes, however, and despite their aforementioned history, are not likely to align themselves with science, nor are scientists likely to perceive much of their methods (or "prestige") in journalism.

Panels at the many conferences, workshops, and events I attended present Durkheimian moments of "conflict and contact" and "ritual solidarity" where an articulation of norms arises in order to situate what it is that journalists should and shouldn't be doing in relation to climate change. Similarly, journalism is under enormous pressure as an industry *and* a profession as a result of new media disruptions to professionalization, influence and authority, and business models. These disruptions have acted to transform the journalist-audience relationship such that audiences are now users who sample from a variety of sources, and spreadability, as Henry Jenkins argues, acts as the primary influence as opposed to the previous era of broadcast hegemony.[16] Thus the Weberian elements are present, too, as "the newsroom" disperses, is altered, and/or becomes irrelevant and hence institutional control and pedagogical inculcation are both in decline. Much like the historical conditions Schudson describes, professionals since the early 1990s have been rallying to explain, dictate, and valorize professional journalistic norms in the face of profound changes in the news industry. Yet this same era has witnessed the rise of citizen journalism and the means to challenge journalists. Some have gone as far as to call this "the golden age of fact-checking," noting that journalists now have their own "peer review" mechanism.

Trust has long been seen as the most important relational aspect for a journalist and his/her audience. It's trust that's seen as keeping audiences tuned into Brian Williams or subscribing to the *New York Times* and *Washington Post*. But Schudson argues that "decreasing levels of public trust in news [are] not so much a matter for alarm as an index of epistemological shift—that people in recent years no longer view journalism just as an institutional provider of informational content, but as an epistemological performance or process of knowledge production"[17] (2013, 196).

Jenkins and others who work on fan culture and new media would argue that the public has always been savvy in their responses to broadcast and institutional forms of knowledge production—that they have always been "making do," responding, remaking, and rereading their own bricolage. This process and an interacting public are only becoming more evident, visible, and possible as media changes beget a plethora of news and information sources, platforms, and devices that are increasingly used to produce and share as much as they are to read/view what's available. Yet the shift exemplified through blogging and other platforms marks a definite regime change in terms of both transparency and accountability. In her comparison of norms among bloggers and professional journalists, Jane Singer sees the work of bloggers as partly being about letting journalists know when they haven't lived up to journalistic norms. Other scholars like Florian Sauvageau contend that journalistic forms and styles must be rethought, as "journalism, previously a lecture, has now become a seminar or a conversation" (2012, 40).

Since I finished fieldwork and began teaching in a journalism school, issues related to transparency, verification, and expertise have only escalated, as have forums in which to discuss the ways that new media challenge journalistic forms, styles, and assertions. As I noted in the introduction, one journalist referred to reporting on climate change as akin to "parking your car under a bunch of starlings." In other words, even when observing a high standard of journalistic practice with regard to fact-checking and adjudication of claims and evidence, journalists who report on climate change are likely to receive attacks from all sides. At a recent conference dedicated to looking at online communication of climate change, Revkin titled his keynote "Is the Internet Good for the Climate?" and then wrote a blog post that also captured Twitter responses to his provocation. His conclusion is that transparency and multiple audiences are now the norm, and journalists have little choice but to "embrace it." But clearly this isn't an even response. Late in 2013, *Popular Science* took the extraordinary step of shutting down its comment section, explaining that the debaters and naysayers were too numerous and "nasty," and citing research by Dominque Brossard and Dietram Scheufele[18] that found online readers were influenced by uncivil comments to view issues as falsely polarized (LaBarre 2013).

Navigating expertise in the midst of ongoing, evolving research presents a particularly daunting task, then, as audiences are asked to trust the conclusions *and* ethical questions offered by scientists and the journalists who rely on them for their expertise. The array of political, advocacy, and policy-oriented groups as well as social movements present another set of challenges both in terms of their role in the ongoing saga of climate activism and policy development and the immediate feedback now available through online media outlets. New media has put ethics front and center for many who work in media, but climate change, more than many ongoing issues, demonstrates aptly that the role of journalist as educator, informer, or advocate is up for debate as "new antagonisms open up between those who produce risk definitions and those who consume them" (Beck 1992, 46). I want to turn now to a story that Revkin did before he left his position in 2009 as a reporter for the *New York Times*. It marks one of the earliest instances of what I am describing here and perhaps explains the title of his more recent talk and its conclusions.

Not the "Shrill Voices Crying Doom":
Blogging, Alarmism, and the Middle

No one reporter has been more closely and consistently linked to the coverage of climate change during the past two decades than Revkin. While at the *New York Times*, Andrew Revkin reported on a prodigious number of stories related to climate change—by his own count, hundreds. Almost everyone I encountered during the course of my research considered his reporting a primary exemplar in terms of its quality, reach, influence, and longevity. If there were a category for a widely acclaimed "expert" on *climate change reporting*, Revkin would be at the top of the list. Not that he doesn't have his detractors on both the left (progressive) and the right (skeptic) ends of the spectrum—respectively, people who either think he isn't "blowing the whistle" hard and long enough to effect massive political and personal change, and those who think he has it all wrong and is part of a vast conspiracy to misinform and defraud the American public. But then climate change reporting tends to attract passionate responses and criticism. Such is the ethical nature of the issue and the difficulty of reporting on it.

With polling numbers on the upswing and a crush of media attention pointed toward climate change in mid-2007, I asked Revkin where he thought the state of climate reporting was at that point. His reply lacked much of the optimism I had heard from many at this point in time.

> The media went from the tendency of ignoring it all together through that stage of equivocation where they just use the old media template of the balance template, yes person and a no person. So now, it's just— it's almost oversimplified because we know the basics, you know, more CO_2 equals warmer world that means we know everything with equal confidence. And anyone who looks carefully at science knows that's not the case. The things that matter most to society are the least certain, whether it's the pace of sea level rise or where, regionally, you're going to have the worst outcomes, or what's going to happen with hurricanes.

To be able to navigate this uncertainty, climate change reporting requires that reporters develop a familiarity with several fields of climate science, differing climate models, and different methods and schools of thought in the field of economics. Reporters should also be at least somewhat aware of the various kinds of multilevel, often global institutions and advocacy groups at work on climate issues in order to understand the behemoth of decades of public debate within which their stories may well circulate. New media has added another dimension to this as well. There is a dedicated section of the "blogosphere" that is alive and well to most climate stories and responds vociferously with occasional support but more often with blistering critique. Reporters tend to get the worst end of crowdsourcing when a contentious issue is at stake.

In 2007, Revkin began replying to and intervening in the blogosphere with his own blog, *Dot Earth,* on the *New York Times* website. But prior to that, Revkin published an article on January 1, 2007, in the *Times* that captures the next evolution of the difficulties and stakes of reporting on climate change. Headlined "A New Middle Stance Emerges in Debate over Climate," the article reported on "some usually staid climate scientists in the usually invisible middle" who were speaking up "amid the shouting lately about whether global warming is a human-caused catastrophe or hoax." In trying to establish evidence of "the new middle" among climate scientists, Revkin quotes MIT scientist Carl Wunsch as saying: "Climate change presents a very real risk. . . . It seems worth a very large premium to insure ourselves against the most catastrophic scenarios. Denying the risk seems utterly stupid. Claiming we can calculate the probabilities with any degree of skill seems equally stupid."

The debate then is not over whether or not climate change is an issue or poses a problem with society; rather, it's about how to talk about it and build "public support." It's about the "appropriate response" to the facts—the ethical dimensions and meaning of scientific findings. Revkin later quotes Mike Hulme from the UK as saying that he found himself "increasingly chastised by climate change campaigners when my public statements and lectures on climate change have not satisfied their thirst for environmental drama."

Hulme's fear was that "the discourse on catastrophe is in danger of tipping society onto a negative, depressive, and reactionary trajectory." Hulme and Wunsch are operating in two very different national media environments. And while it is easily argued that the internationalization of climate research and advocacy as well as the global 24/7 media market-place have rapidly connected these environments, the political and polling responses to climate change couldn't be more different in each context. The UK has managed to consistently rank higher in public opinion on climate change than the United States, and the UK has a government that has responded with policy changes related to climate change, unlike the U.S. at that time.

Despite these differences, Hulme, Wunsch, and several other U.S. scientists including Roger Pielke, an active political scientist on climate issues, consider themselves apart from what Revkin characterizes as the "shrill voices crying doom [that] could paralyze instead of inspire." Pielke's term for this "middle" group is "nonskeptical heretics." It's a confusing term and one that requires some grounding in the scientific and policy debates, as well as the debates over how to explain climate change to the general public. The nonskeptical half of the term refers to the fact that these scientists in "the middle" are not skeptical of the scientific facts related to climate change, but they are *heretical* because they are unwilling to go along with the strong urges to advocate vigorously for immediate change in the face of the likely catastrophe predicted by various climate models.

The worry, Revkin explains, is that Gore's now very popular film acts to alarm the public and yet doesn't go far enough in proposing adequate responses. He paraphrases Jerry Mahlman, a climate scientist at NCAR in Boulder, Colorado, as saying that climate change needs to be treated as "a risk to be reduced" rather than "a problem to be solved." In contrast, James Hansen and John Holdren, well-known climate scientists, were quoted as those who "say there is no time for nuance" and that "moderation in a message is likely to be misread as satisfaction with the pace of change." That scientists would be discussing discursive strategies for engaging the public and that it would be considered "news" reflect the tenor of this moment in public climate change discussions and news coverage. What's more is that the presentation of facts presents a view on the urgency of those facts.

The response in the blogosphere provides another facet that registers in part the hybridity of the digital-traditional discussion and coverage of this issue. Almost immediately after Revkin's article was published, Patrick Kennedy at the *Daily Kos*, Roger Pielke, and Carl Pope, founder of Sierra and *Huffington Post* blogger, took up Revkin's article, generating discussion and responses online. While quick to defend the veracity of Gore's film with minor exceptions based on his own research of scientists' responses, Kennedy thought Revkin did a good job of reporting on the "real debate."

> With the new Democratic Congress and the cooperation of the mainstream media, the phony debate, with climate scientists on one side and the [Senator] Inhofes of the world on the other side, will, with luck, disappear in 2007. The real debate, not limited to climate scientists, is about what is the best way to engage the public and policy makers on the serious challenge we face from global warming and move forward.

Kennedy's reference to "the Inhofes of the world" is an allusion to skeptics like the senator from Oklahoma who called climate change "the greatest hoax ever perpetrated on the American people."[19] Responses to Kennedy's blog post ranged from support for Revkin to disgust that he was attacking Al Gore and suggestions that the *New York Times* did not want to see anything done on climate change.

Carl Pope on the *Huffington Post* (HP) headlined his post "Why Media Doesn't Get It." He argues, "The American media needs to cover global warming as the urgent real-action-required-now challenge that it is." He lists a recent NBC story as well as Revkin's story as evidence that the American media are not covering climate change as such. Though Pope calls Revkin "one of the best writers in the most respected paper," he took issue with the fact that Revkin's story reinforced "the misleading notion that there remains a serious scientific debate about whether or not we need to take action." Again, that scientists would be debating action and response and not the veracity of findings and methods speaks volumes about the ways in which scientists have been drawn into a "debate" of some kind about things decidedly nonscientific and unrelated to their expertise. Pope's fear was that if the media didn't "get it," then the public wouldn't either, in enough time to make a difference. What lies perhaps at the root of such fears is the need for ways to integrate a spectrum of information about risk into institutionalized mechanisms for communicating and addressing those risks. This spectrum of risk is an essential characteristic of climate change as an emergent form of life that, as Pope signals, exists on ethical terrain (Fischer 2003).

Revkin responded on an interim blog he kept at Amazon.com before launching *Dot Earth* on the *New York Times* website. He initially noted that the piece was "generating quite a few sparks," citing Kennedy and Pielke's blog. And he said that one "veteran climate scientist" had sent out a mass e-mail saying Revkin had "done a 'great disservice' by writing it [the article] and concluded 'shame on you.'" Revkin directly addressed Carl Pope and copied the response he had posted on Pope's blog.

> While it may be old news to Carl and many Huffington readers that virtually all serious scientists agree that more CO_2 will make the world warmer (thanks in part, hopefully, to my 20 years of coverage), this does not mean most Americans have absorbed this point yet. There are tens of millions of disengaged or doubtful or simply uninformed people out there, many of whom shy away from loud voices. For them, the public discourse is largely (and incorrectly) a big Fox-style debate. My goal was to point out that *even* the normally invisible middle in climate science sees human-forced warming as dangerous and requiring a prompt response.

Revkin went on to argue that those in the middle don't want to sound alarmist, but they are not skeptics in any way, nor do they offer comfort to those who downplay the role of humans or the lack of need for action. His main point is that the middle shouldn't be left out of debates about "how best to limit climate risks in a human-warmed world." These "middle" scientists, in Revkin's characterization, are trying to set a different frame for discussion about how to limit risk.

When I asked Revkin about this exchange with Pope and the story he wrote, I prefaced my question by saying that I, too, had heard remarks about alarmism in Gore's film from scientists I had interviewed. I said that I had also heard several observe that skeptics were (now) saying that climate change may be happening, but not to the degree Gore dramatizes in the film. Revkin responded this way:

> The more one side tries to work hard to motivate people around the idea that we face a climate crisis that requires urgent action, the more that it can almost empower those saying it's all a hoax if the crisis is defined by oversimplifying the phenomena. Because then it leaves you open to criticism that you are not being careful with science. Instead of saying, yeah, it's a crisis, but it's on a century scale, and the worst impacts are going to face generations yet unborn, which is most likely the reality and makes it much harder to sell. But, at the same time, it is true to the facts. So there's this dynamic in this issue that drives it to the edges because everyone hates the middle, which is where we know the most. The middle is gray, and in our current political dynamic, gray doesn't really work.

That there's something between crisis and dissent does not generally make news, and environmental policy is often driven by major crises—Alar on apples, the *Exxon Valdez* oil spill, and the Cuyahoga River catching fire because it was so loaded with polluting toxins and chemicals, to name a few.[20] Climate change instead presents questions of ethics about how to deal with risk, the nature of that risk, and what kinds of institutions and assemblages should be created, mutated, and destroyed in order to address a future with risk (Fischer 2003; Fischer 2009). How much do we want to leave for future generations to deal with? How much do we want to hope that the risks inherent in predictions related to a warming climate aren't all that bad?

Journalists are tasked with articulating the ongoing societal relationship with notions of and futures with risk. Ulrich Beck (1992, 2002) has theorized that the current epoch is marked by a transformation from modern industrialization to a risk society, marked and marred by unintended and unpredictable consequences. In these terms, predictions related to climate change illustrate that industrialization has created human comforts and widespread urban living, as well as visible, felt instability and chaos at the Earth's poles that will filter downward/upward to the industrial infrastructure that spawned such chaos at some point in the near and/or distant future. This disconnect between cause, consequence, and the conditions that make decisions that cause such consequences possible defines the risk society such that, Beck argues, "its heart rests in the mass media, politics, and bureaucracy—not necessarily at the site of its happening" (2002, 4). This is the disconnect that vernaculars bring into sharp focus—that the discourse at the level of policy and media is not always recognizable on the front lines whether they're in rural Alaska or rural Oregon. Local reporting provides a site of tension and a clash between seemingly disparate forms of life—a moment where observations of causes and effects on the ground talk past each other, making it easy to deny, ignore, or rage against the claims and/or priorities of each other.

Despite the power afforded to media to shape discourse, mainstream media has struggled with its own set of negotiations as climate change has developed as a news story, a scientific fact rife with uncertainty and a wide spectrum of possible outcomes, and an issue for advocacy. Every journalist I spoke with or heard speak on numerous panels and at workshops can cite multiple instances of such challenges, and many have developed a point of view about how and what has gone wrong and right with reporting on the issue. Beck notes that risk society could also be considered the "science, media, and information society" where debates and struggles occur over how to define risk and its degree, scale, and urgency. "The middle" that Revkin identifies is exactly this kind of problem, buffeted as it is by those on either side who make claims to what must or must not be done in order to prepare for the immediate or distant future of risk.

Information via media, and in particular mainstream media, is seen as essential to the workings of democracy. Schudson (1998) has argued that the informed citizen that undergirds dominant democratic ideals is rapidly being reformed into a monitorial citizen with access to multiple and continuous streams of information. The journalist then is caught between demarcating the outlines of risk, multiple forms of life that all seek to define climate change in various ways, and traditional notions and obligations associated with professional norms. And in the background remains the "starling effect"—the robust questioning and counterclaims from those who recognize climate change as a problem with ethical and moral contours and those who do not.

What these more iterative and interactive processes confront is that media is most often only able to deliver a version (or versions) of the truth: a professionally coded point of view or an interpretation of "what is really happening." And while there is the arresting headline to create, there is a tacit acknowledgment of such interpretations building up over time and space. Yet professionalism and the system by which news is produced have created enormous barriers for others to add to dominant interpretations and conclusions. This is in part what the confrontation with new, social, and multiple media entails—and what makes blogging and social media such a rich, multilayered set of interventions. The increasing circulation of other narratives has laid bare the notion that facts get constructed, produced, and socialized. It has in many ways opened up the epistemological aspect of claims as well as the ethical component—and shown how they are, despite best efforts at objectivity, inherently linked and particularly so on an issue like climate change where facts must and do lead to larger moral and ethical questions.[21]

In 2010, Revkin's *Dot Earth* blog moved from the news section to the opinion section of the *New York Times* when he took a buyout package in late 2009 and ceased to be a staff reporter. Some reports like that by Bud Ward pointed out that Revkin was exhausted by the pace of 24/7 reporting as well as by the major furor caused by several articles he had written—one of which is the "middle" story I detail here. Revkin remained relatively quiet about those factors on his blog. When it moved to the opinion section, he said that such a move allows him "to say what I think in ways I could not when I was a *Times* reporter" (2010). Does this make him an advocate? Some of the comments to this blog post titled *Dot Earth 2.0* lamented the end of what they felt was a last bastion of real debate and discussion, where both (or many) sides of arguments were well represented. Revkin answered this by saying: "Don't expect momentous changes. I'm not going to suddenly be revealed as an ardent liberal or conservative. I am an advocate, for sure—for reality."

This is likely a statement many journalists would agree with, and it speaks to the ethical obligations inherent in reporting. Yet what form of life composes which spectrum of reality—in other words, reality for whom, and by whom? And could advocating for "reality," however defined, constitute a form of near-advocacy?

Conclusion

Climate news and feature stories that end up in major media sources are now subject to immense scrutiny, criticism, and counterclaims from concerned audiences with diverse perspectives and vested stakes and the means and channels in which to respond. This is what a plethora of new sources online facilitate, and it is in part what makes covering climate change a more challenging task than it's ever been. For those engaged in this issue, how evidence is deployed, who is speaking for it, and where scientific knowledge has been produced are vital details. The role imagined for journalism in our democracy has traditionally been one of informer, agenda setter, and watchdog. This function of forum provider, chief discussant, and extant verifier is still a very new one for journalists and news providers to navigate or even to understand.[22]

Revkin's career trajectory illustrates the ways in which notions of audience and journalism are changing. He is among an elite and small cadre of science journalists who have shaped media conversations about climate change—indeed, it is arguable that Revkin has established a standard for journalistic articulations of it. Yet even before he left his full-time reporting role, he was experimenting with the ways that blogging and social media opened up new avenues for interaction and experimentation. What blogging makes possible and evident is a tracking of the minute shifts and ways in which climate as a form of life is continually expanding and contracting. Blogging provides a way for direct public response to what journalists like Revkin are reporting on, and bloggers create audiences of their own that continue to debate and respond to what is being reported on and what blog commentaries are being offered on the reporting. As media technology expands, these conversations and interactions will only expand and transform existing media platforms (Domingo and Heinonen 2008; Hermida 2011; Jenkins 2006a; Jenkins 2006b; Jenkins and Thorburn 2004; Usher 2010). Emerging conversations on and via Twitter (a micro-blogging platform) that point back to longer blog responses and/or facilitate discussions about conference presentations, news articles, or recently published research provide a current example.[23]

Beck's 1992 observations about our "risk society" are turning out to be extremely prescient, then, when applied to climate change reporting—regarding journalism's role as articulators, cries of alarmism (or indifference), and the increasing antagonism between producers and consumers of risk. The late Stephen Schneider put it aptly when he told the reporters in Oregon that "all good science . . . gives you probability distributions"—in other words, a range of risks that produce a range of unevenly distributed effects and potential scenarios that benefit some and devastate others. Yet scientific consensus as it is reproduced in media often elides a wide spectrum of risks in favor of generating a unifying message so the public is not confused; this is what Revkin's articulation of a "middle" pushes against. Wunsch, in the article, even takes aim at the notion that scientific evidence can produce reliable probability distributions, yet this has been seen as a key application of climate change for public and policy engagement.[24] Charges of alarmism are levied by skeptics at anyone who acknowledges the ill effects of a future with climate change, but it's also applied to those unable to articulate adequately just what a spectrum of outcomes might entail.

Climate change reporting has begun to shift toward thinking through what to do about climate change, and it is in this sense that reporters must navigate a stance with regard to near-advocacy. Facts are not settled, nor is it certain exactly how findings and modeling might evolve in relation to new findings and models. So what or whose truth and when are particularly key questions when it comes to assessing a future with, *and* reporting on, climate change. In this sense, journalists must contribute to articulations of climate change's form of life: what it means, how to speak about it, what knowledge is relevant, and why it matters. Journalists also must struggle with the evolving nature of that form of life and the rules and grammars that are prescribed by scientists, advocates, and political figures of all stripes. Climate change as a form of life is a contrast to considering it a stable entity or suite of facts that journalists are having a hard time explaining to the public so that they either care enough to do something about it or "understand" the science. Climate change, as an evolving form of life, *demands* that journalists rethink professional norms and practices, particularly around gatekeeping, objectivity, conceptions of "the public," and what's deemed "relevant."[25] Near-advocacy thus operates on a spectrum that recognizes both the collective and individual efforts of journalists to negotiate and develop their own relationship-building efforts with "the facts."

At the SEJ conference in 2008, one reporter claimed that all climate stories were already being funneled to policy and political reporters at their media outlet—in other words, that it had ceased to be a "science story." But the definition of climate change as a form of life remains in flux, and increasingly as blogs and other forms of media intervene to shift the reception of original reporting, this resistance is likely to produce more hybridity, expertise morphing, and debate for those with a stake in how risk is accounted for. Blogging and other means of talking back to journalists and weighing in on public discussions also have the effect of making journalistic norms evident and holding journalists to account (see Singer 2005). Ethical or good journalism is seen to be independent, unbiased, and objective, and when it is seen to do or be otherwise, the outcry from active audiences and commentators is deafening for those with profile and reach like Revkin.

Finally, media change is altering not only how media are produced and consumed but also the role of information in society. Whether and how the public comes to be engaged with climate change is determined not only by access to information about it, but by the ways in which it becomes meaningful—by the form of life it assumes. Journalists are messengers whose ability to invest meaning, ethics, and morality are limited by professional norms of nonadvocacy, near-objectivity, balance, and accuracy. Climate change sounds different within communities formed through belief, identity, and shared values than it does coming from journalists who are usually hoping to reach the widest possible audience. Accounting for this difference and the vernaculars therein provides a challenge for societal expectations of actively engaged citizens and the delineation of civic epistemologies—for the way that scientific evidence and findings come to matter for a society.

CHAPTER THREE

Blessing the Facts

In early 2008, I traveled south to Orlando, Florida, or what felt like the furthest place in the world from the Arctic, MIT, and the Northeast corridor of the United States, where I had done the majority of my ethnographic research to date. Known worldwide for its many tourist attractions and theme parks, Orlando is also home to Northland Church. Northland is a megachurch that claims twelve thousand members worldwide who either attend services in Longwood, a suburb of Orlando, or log on via the Internet. Traversing a network of freeways and major multilane routes, I fumbled my way to Northland's sprawling campus in order to attend the 2008 Creation Care conference. Set in a district where extra-large parking lots are not out of the ordinary, Northland was originally the site of a skating rink and sits across the street from a dog kennel and race track. The church's setting is not at all what one might expect in times past of a stalwart church presence in the community. In other words, a traditional white-steepled fixture on a tree-lined residential street it was not. Instead, Northland is part of the sprawl of contemporary suburban and exurban landscapes.

The parking lots were mostly empty on the day of the conference, and we didn't meet in the sanctuary that holds over 3,000 people on Sunday mornings. Instead, we met in a side building more appropriate for the approximately 100–150 people who attended the daylong conference. When I got to the door of the building, there was a small lineup to get in. As in many small churches on Sunday morning, Northland's senior pastor, Joel Hunter, was greeting each person individually as they came in the door for the conference. Waiting in line, I listened as Hunter acknowledged the heavy-set, white-haired man in front of me and his two companions. When Hunter heard the man's name, he was elated and repeated the man's name loudly so I could hear it. It was a name I recognized as well. It turned out the man was a pastor and a prolific author whose many books Hunter had read and enjoyed.

When I greeted Hunter next, I identified myself as a researcher from MIT interested in how climate change was being communicated for Americans. He was elated and said, "We need that!" I was surprised by the openness, but as I was to discover, this characteristic went hand in hand with the ecumenical order of the day. Hunter later hosted a panel that sounded like the beginning of a bad joke with a priest, a rabbi, an imam, and a pastor. The panel, however surprising, was devoted to understanding how it is that other religions made sense of environmental concerns. In this panel and throughout the day, there was a continual acknowledgment of both the renegade nature of such a meeting and the need to translate climate change so that evangelicals recognized it as "their" issue and one that required action.

At lunch, I found myself a seat at the same table as the prominent evangelical author/pastor I had followed in the door. I was intrigued by his presence there, particularly because he was an attendee, not a speaker, despite his accomplishments. Lunch was informal, intended to help the relatively small number of attendees network with one another. We were surrounded by hastily assembled booths from Christian publishers with a surprising number of new books on Creation Care–related topics and Christian-affiliated conservation groups like A Rocha. Our lunch table of about ten was a motley crew of students, Christian lay workers, and this pastor. I struck up a conversation with the young woman beside me and discovered that she was the pastor's daughter and had just finished her undergraduate degree.

I gave her a more in-depth introduction than I had given Hunter in our quick greeting. I told her I was conducting research on the diverse ways climate change is being communicated to Americans and that Creation Care was one group whose efforts I was looking at. She was enthusiastic in her response. She had just finished working with and writing a report for a secular environmental group to try and help them understand how to reach out to evangelicals. I was surprised that this kind of initiative was under way, but she said that environmental groups were starting to realize that many of their members were believers of some kind.

Her personal story was equally compelling and surprising. She told me that she didn't grow up in a home that was concerned about the environment. While pursuing her undergraduate degree, she had spent a year abroad and had come back converted to concern about the environment. She didn't specify what it was that caused her conversion, but she did note that her newfound priority was a sore point with her father.

It was at this point that her father joined our conversation. It turned out that she was his eldest of several children, and he was here partly out of concern that had begun with her turn to environmental issues. He nodded when she said it was a difficult thing for them to discuss for quite a while, but clearly, owing to his presence at this conference, they had found some common ground. So I asked him about how Christians were talking and thinking about climate change and becoming convinced of the need to act. After a thoughtful pause, he said that Christians were "skeptical" of science—going back a hundred years, they viewed science as "suspect." He said that science can't be the reason to act. The argument and appeal for evangelicals has to be on "moral" grounds. It has to be about "stewardship."

This statement was reinforced in the conference speeches, a growing number of books on Creation Care,[1] and interviews I've since conducted. Yet it was stunning to me at the time both for its clarity and for the questions it posed for informing Americans about climate change. For if climate change is not a matter of the public understanding of science, then how is it being communicated by, to, and for this group? What kind of an issue is it for those who are not drawn in by scientific evidence? What kind of language is left when science is not the primary tool for presenting the issue and its implications? These are variations on the questions I have posed with each group I've researched for this book, but they are perhaps more poignant here because of the flat rejection of science as the *sole* basis for evidence upon which to become persuaded of the fact of climate change and the duty to act.

Creation Care, an emergent and multivocal social movement, was begun specifically to address this by taking climate change out of the realm of science and environmental activism and situating it instead as a moral issue understood through evangelical teachings about the Bible. It's as a direct result of that process that individual and collective responsibilities to care for the environment become apparent. Christian scholar and novelist C. S. Lewis perhaps said it most eloquently in his 1945 speech "Is Theology Poetry?" In the speech read to the Oxford Socratic Club, he wrestled with "scientific cosmology" as a view of the world based on rational observations, and he explained why he opted for the Christian theology—not because scientific methods can't tell a lot about the world but because they limit how and how much one sees the world. He ended the speech this way: "Christian theology can fit in science, art, morality, and the sub-Christian religions. The scientific point of view cannot fit in any of these things, not even science itself. I believe in Christianity as I believe that the Sun has risen, not only because I see it but because by it I see everything else."[2]

Lewis still holds an enormous amount of influence in evangelical circles, particularly because of the ways in which he sought to make sense of Christian morals, ethics, and teachings. And, in this speech, he manages to draft a contrast that still stands for many—where science, despite its truths, is found inadequate and ideological even as it claims an immanent terrain of "Reason." So how exactly does a scientific issue become a Christian one, with moral weight such that actions are required as a result of the science as understood through a Christian framework? What counts as evidence?

The tendency in academic and scientific circles has been to consider such frameworks for interpretation as "ideology," and in doing so there is an inherent dismissal of the evangelical commitment to truth-seeking and, I would argue, a missed opportunity for symmetrical analysis of how evidence comes to matter. Here I want to move evangelical responses to climate change onto an epistemological terrain. Christian theology presents a way of knowing and apprehending the world, as well as a set of norms for how individuals should act in the world. In this sense, there are some similarities (and key differences) to the ways in which it acts as a contrasting knowledge system to scientific knowledge much like an indigenous knowledge system, and the moral prescriptions or norms that flow from how evangelicals see their role in society much like journalists. In both cases, expertise and interpretive practices are distinct and valued. This book as a whole looks to locate these similarities with regard to climate change and to recognize where epistemological differences matter and why—such that the emergent aspects of climate change's forms of life become evident. Resistance to climate change and recognizing where it presents a challenge to norms and epistemologies set up a very different set of problematics around public engagement, who gets to speak for and about the issue (expertise), how credibility is established, and what those articulations sound like. Climate change sounds different coming from the pulpit on Sunday morning than it does elsewhere. In these sites, there is a complicated interplay between vernaculars within which climate change is struggling to gain a foothold.

This chapter dives into these questions by tracing threads of expertise, history, discourse, and vernacular. It draws in part on Susan Harding's seminal work, *The Book of Jerry Falwell* (2000), where Harding immerses herself in evangelical discourse—making evangelical language "her field site"—and formulates the notion of the group's vernacular: "to show how Bible-based language persuades and produces effects" (xii). Discourse and vernacular expression, as Harding points out, are an essential aspect of what it means to be evangelical in America—pastors are public figures "who expect that their words will be studied and discussed." This research, however, is not expressly focused on the way vernacular circulates within evangelical communities, but rather on how vernaculars act as a bridge to/from evangelical communities and beliefs: between science and evangelical thought, between evangelical activists and media and policy arenas. Such bridging and movement through and between vernaculars is a process I am calling "translation," and it enrolls an assemblage of institutions, materiality, and modes of speech in order to form articulations of climate change that resonate with evangelicals (Fischer 2003; Foucault 1995, 2003).

One of the primary arguments put forward by Creation Care leaders I interviewed is that who is speaking matters as much as what they say. Christian leaders (and a few select Christians who are also leading scientists) must "bless the facts" in order for them to have traction and resonance within Christian communities. The notion of "blessing facts" neatly encapsulates the ways in which climate change is being cast as simultaneously intellectual, scientific, and moral, and it speaks to the weight given to vernacular within evangelical communities. It also, however, glosses over or even dodges the traditional debates over evolution that have pitted science against evangelical beliefs while directly confronting those who have chosen to side with climate skeptics. Creation Care translates climate change primarily into a biblically mandated concern for the poor—for how scientific predictions will exacerbate the afflictions of those less fortunate worldwide, as well as harkening back to older conceptions of biblical stewardship or "tending the garden," referencing the idea of the natural world's beginning as the biblical Garden of Eden. Climate change thus provides an opportunity to reinforce norms about how Christians should respond to issues of inequality and poverty while eliding any critique of industrial capitalism, race, and class issues in America or recent globalization.

It takes this work of navigating stakes for the group and finding ways to articulate within their vernaculars and epistemological frameworks to make climate change the group's concern. Particular to Creation Care, however, has been the uphill struggle of dealing with historic public debacles and debates with scientists and scientific institutions about evolution—both recently with the 2005 *Kitzmiller v. Dover School District* decision and "a hundred years back" with the 1925 Scopes trial (Harding 1991; PBS *Nova* 2007). As well, Creation Care must confront political alignment and ties with the Republican Party and leading evangelicals who are not in agreement with climate change and collaborate with others who are expressly skeptical of climate change (and not just science in general). This chapter records and narrates what might be summed up as "frictions" intrinsic in this constantly evolving process of translation (Benjamin 1968; Fischer 2009; Harding 1991, 2000; Povinelli 2001; Tsing 2005).

Creation Care can also be seen as one way in which evangelicals are moving into a differently conceived and politicized notion of which core issues should concern civic-minded and active evangelicals. As well, because of the involvement of figures like Joel Hunter, these changes can be viewed as a part of larger shifts regarding the structures and technologies inherent to evangelical church attendance and organization.[3] Climate change then as explained within evangelical discourse is one of many issues embroiled in larger changes instigated by new leadership, technologies, and demographics.[4] This chapter thus provides an account of relationship-building with the scientific facts and a negotiation with collective historical memory, ideology, and epistemology. Set within the larger framework of this book, how evangelicals contribute to and challenge climate change further evidences the issue as an emergent form of life whose meaning, definition, rules, and grammars are very much contested, as are the kinds of knowledges and interpretive frameworks associated with it.

What It Sounds Like Coming from the Pulpit

I wasn't sure what to expect at the Creation Care conference when I arrived. I was surprised and intrigued by its intimate size. I had struggled to make inroads via e-mail with many of the people I had identified as key leaders and was anticipating the insight this event might provide. The event opened first with a prayer by Joel Hunter where he acknowledged: "We are the receivers of your [God's] great creation, and we confess that we have not treated it with the utmost respect . . . and we want to do better." This was followed quickly by a short drama sketch by two members of Northland in order to "set the context" for the forthcoming session.

The sketch began with a woman sorting out things outside her house and doing a "green audit." A man arrives and tries to guess in a humorous way whether she has gone Buddhist, vegetarian, organic, vegan (neither were sure what that meant, which elicited a small laugh from the audience), Adventist (which got a big laugh), or something else. The "green" individual responded that no, she wasn't any different and still went to the same church (a refrain she repeated a few times). This "going green" had not affected any other area of her life except to force her to buy more expensive light bulbs (which also got a laugh). The punch line from the other actor was *Why you doing it then?* The scene ended with the voice of Kermit the Frog singing, "It's not that easy being green," which got a round of applause, mixed with laughter.

Hunter came back onstage then and said, "Those of us making this transition are confusing many of those who are trying to put this whole thing into some sort of category." He described the evangelical community as "being late to the table" in terms of taking on the issue of care for the environment, "confusing much of our congregation that has a basic responsibility" to do better on simple things. He said that the conference day was meant to be a "training session" to equip pastors and lay leaders "wherever they are at," implying there were a wide variety of positions on this topic. As if to explain the small numbers, he said that right now Creation Care was a "rather concentrated network of leaders" that is "really expanding very rapidly." He said, "Pastors are very intimidated to address this with their congregations because of those more *radical links* that people have in their minds, and so it's a risk for pastors to bring this up." It wasn't all a difficult road of persuasion, though. He made reference to the surprising response particularly from the younger generation of evangelicals. Hunter introduced the first speaker, Tri Robinson, a pastor of a Boise Vineyard Church,[5] who has rapidly become a leading voice on issues related to evangelicals and the environment. He's written two books, *Saving God's Green Earth* (2006, cowritten with Chatraw) and *Small Footprint, Big Handprint* (2007), both of which lay out a vision that includes the biblical basis behind care for the environment, Robinson's personal experiences, and a blueprint for living an ecologically principled life both collectively (as a church) and personally in the home.

Unlike Hunter, who was in a suit, Robinson wore jeans, cowboy boots, and a bright yellow shirt. When he got up to speak, he said it was the first time he had spoken this message on the east coast, and he said when he arrived in the morning he thought, "Uh-oh—because I was the only guy who looked like he was going fly-fishing. Usually if you see me in a tie, somebody's getting married or somebody just died." Robinson is from Idaho, and as he details in *Saving*, he raised his now grown children mostly off the grid on his family's ranch near Boise while pastoring. Prior to becoming a pastor, he had earned a degree in biology and ecology and had been a teacher for twelve years.

Robinson said when he decided to speak to his church about the environment and stewardship, he "was scared to death." He explained that he prepared for six months and formed a task force within the church and discovered that many of his parishioners worked in conservation, fish and game, parks and recreation, soil conservation, and national forest service. "I had all these undercover, closet conservationists and environmentalists afraid to admit in church they were environmentalists. And, on the other hand, in their workplace, they were afraid to admit they were a Christian. So they were living these double lives, and in some ways, I was, too."

He explained to the conference crowd that, in addition to his science degree, he came out of the Jesus movement in the 1960s that embraced radical environmental values, but then "the thesis that emerged was it's all gonna burn anyways," so those concerns became a part of his "old life" that he left behind. He equated environmental concern with other elements that he didn't like about his life before conversion, and so he said he decided, like others who had come out of the 1960s, that they would just preach the gospel and that would be their life.

Yet Robinson pointed out that part of the Great Commission to go and preach the gospel included "tending the garden and caring for Creation. It's very clear in the scripture." And when evangelicals dropped this issue, others picked it up.

> We didn't quite like how they [environmental groups or environmentalists in general] did it. Two camps, I think, were formed, and we isolated ourselves from each other. Most of us were really pushing in the pro-life camp and saw that the environmental movement was really a pro-choice camp because the thought came that the problem with this world is that it's overpopulated. What they said was that, really, what we need to do is rid ourselves of unwanted people. We really pushed back [from that].

Not incidentally, this is a concern I've heard many times informally, particularly from a high-ranking representative of the influential parachurch group Focus on the Family (founded by James Dobson) as a rationale for ignoring climate change as an issue. At a public event at MIT, this representative, when asked, said that Focus could not embrace climate change because of the issue's association with abortion-on-demand and overpopulation. These are likely part of the "radical links" Hunter had in mind when he opened the conference.

Robinson referred to his home state, Idaho, as a "red state" and noted he had Republican politicians in his congregation, but he felt confident going forward because he had discovered scriptural principles for his message about the environment. When he gave his first sermon— the one he prepared six months for—he said it wasn't his best sermon, but at the end, he received his first standing ovation in his twenty-five to thirty years of preaching. That sermon, he noted, is still being heard online.

Robinson described himself as being only two and a half years down the road of thinking through and about environmental stewardship in relation to Christian responsibility. He said that he had shown *The Great Warming* at his church and advised others not to do it if they were new to the topic. He got a big laugh when he said he "had to 'fellowship' a few people afterward"—meaning everyone wasn't happy about the film. He said it wasn't "the Al Gore version," and therefore it was more palatable to evangelicals, and he specifically noted the interview excerpt in the film with Richard Cizik, who was in the audience. Robinson recalled what Cizik said in the film—that evangelicals never hear this kind of thinking from the pulpit. He followed up on this point by stating: "Christians are waiting for their leaders to give them permission to care about creation and say that it's okay." This theme was repeated throughout the day.

Robinson went on to provide scriptural references and interpretation for his newfound position on environmental care. He noted that "the problem of people worshipping the creation and not the Creator," which is often expressed as an issue of concern in evangelical circles, is something he's never witnessed in his nearly three years of working on this issue. Instead, he said that it opened new doors for accessing the "unchurched." He's been asked to speak to the conservation league and to partner with the University of Idaho. He emphasized the evangelical duty to proclaim, demonstrate, and participate, and he explained this in several ways: as God being revealed through creation, as a responsibility to address the needs of the poor and "a world in crisis," and as a call to live a life of adventure. Robinson ended by showing a short video of his church's efforts, which include a large organic garden on church grounds from which they feed the poor. He strongly emphasized that his church has not been hurt by these new initiatives but rather that it has been blessed by it. And it has provided him with an opportunity to speak with groups like the Idaho Conservation League, which he described as "the most liberal group" in the state. He said he began his speech to them by "repenting" about Christian attitudes and actions regarding the environment—a sentiment he reckoned they had never heard from a Christian leader before.

Robinson's message raises several themes. First, obviously, is the sense that the environment is a predominantly liberal or Democratic matter. Second, Robinson went to great lengths to point out that this is not a new value being overlaid but an old value of "restoring Eden" that was finally being brought forward. So the work is not just that of parsing environmental concern from a liberal set of concerns but also recovering the biblical fidelity inherent in such concerns. And it is up to the pastor to articulate that fidelity and give permission to adjust the priorities of their church and parishioners. But that articulation is not the final word. Robinson laid groundwork socially in order that his message might be received well by at least a segment of his church. This kind of tactical strategy mirrors on a small scale what Creation Care leaders are attempting to do on a much larger scale, working to build bridges and inroads for their message both inside and outside the movement. And they do so, as Hunter hinted, with the younger demographic largely on their side.

After Robinson's talk, Hunter followed up by warning: "You are going to get some pushback. Some of your people listen to talk radio during the day and they think we're the devil." Indeed, after sitting in the conference for the first hour of presentations, I already felt as if I had entered a renegade camp where, despite a belief in the absolute necessity of this work, there was an acknowledgment that such agitation bucked trends and could/would likely upset many. The roots of Republican affiliation run deep for many evangelicals, diverse as they are, and this is a key factor in the difficulty Creation Care has had moving forward, beginning with the Evangelical Climate Initiative. At the center of this movement, then, is a debate not only about science and evangelical Christianity but also about how and why to be civically active.

Antecedents and New Beginnings:
The Evangelical Climate Initiative and Creation Care

Creation Care is not an old movement, nor is it very institutionalized. In fact, it is a submovement within the larger movement of American evangelicals and one that is still largely nascent. But out of the rumblings of a few, beginning in the early 1990s, Creation Care has grown in order to both peak and transform during the period in which I have followed it ethnographically.[6] It has distinct roots in the after effects of the unequivocal statements made in the 2001 IPCC report. John Houghton, chair of IPCC's 2001 Working Group I, is also an evangelical Christian, and he joined with American scientist and evangelical Calvin DeWitt in order to begin a distinctly evangelical dialogue on climate change. Their groups, the John Ray Initiative (Houghton) and Au Sable Institute (DeWitt), organized a conference for Christians at Oxford University in 2002.

It was at Oxford that American evangelical Richard Cizik, vice president of government affairs at the National Association of Evangelicals (NAE), experienced what he calls a *conversion* regarding climate change.[7] His invitation to participate was part of a larger long-term effort to turn evangelical leaders toward environmental concern by a small circle that includes DeWitt, Jim Ball (Evangelical Environmental Network or EEN), Ron Sider (Evangelicals for Social Action), and Bob Seiple (World Vision US until 1998). In 2002 EEN also launched its national awareness campaign, "What Would Jesus Drive?" And its success was one of several factors that informed the strategies developed after the Oxford conference.

Cizik, Ball, and DeWitt attempted to re-create the Oxford experience in the United States in 2004 at a conference held near the headwaters of

Chesapeake Bay at the Sandy Cove Conference Center. Houghton was a keynote speaker, and the conference was sponsored by EEN, NAE, and *Christianity Today*. The conference produced the Sandy Cove Covenant, which laid the foundation for the 2006 Evangelical Climate Initiative (ECI) and efforts "to dialogue with evangelical leaders" about Creation Care. Later that year, the NAE released "For the Health of the Nation: An Evangelical Call to Civic Responsibility," where it listed Creation Care as one of its priorities.

I first encountered Creation Care through the ECI, a declaration signed by a group that included mega-church pastors, Christian college presidents, and para-church organizational and thought leaders. What made it "news" was that even before the release of the declaration, it was met with hostility by politically active, Republican-aligned evangelical leaders like James Dobson, Charles Colson, and Richard Land (Beisner 2007; Blunt 2006; Goodstein 2006, 2007; Vu 2007; Wildmon et al. 2007). ECI was discussed in the *New York Times* before it was officially released because of a letter signed by the latter group of leaders sent in advance of the declaration that attempted to circumvent any appearance that ECI spoke for all evangelicals. This letter spawned the Interfaith Stewardship Alliance and reinvigorated the 1999 Cornwall Declaration, which maintains that the science behind climate change is still "uncertain," a response which I'll examine later in this chapter. Some wondered whether this public disagreement was a major crack in the conservative movement that would reverberate in the political sphere and whether or not it would influence the policy of the Bush administration on climate change. Though evangelicals directly affected Bush's policy on the Sudan, for example, a similar effect did not occur following ECI.

One prominent opponent of the ECI, Pat Robertson, has since publicly changed his mind on climate change. This made news in 2006. In an interview with CBS, Robertson said he was convinced of global warming by that summer's "record-breaking heat," and he said God told him that more storms were coming (Roberts 2006b).[8] Charles Colson and James Dobson, however, remained avowedly skeptical of climate change. A 2009 blog post by Colson expressed concern that climate change was in danger of becoming a religion itself. Belief in climate change according to Colson's formulation constitutes competition with core beliefs in the Bible, and Colson explains that, like the individual he profiles in the blog post, one could lose their job or standing in secular communities by not expressing belief in and support for climate change.

After the flurry of activity and declarations between 2002 and 2006, 2007 and 2008 were relatively quiet as the movement gained momentum. I followed Creation Care most closely during this period. Then in 2009, things transformed drastically. First, Cizik was forced to resign due to his declaration of support for gay civil unions in an interview on December 2, 2008, with Terry Gross on NPR's *Fresh Air* (Goodstein 2008; National Association of Evangelicals 2008b; NPR 2008; Salter 2008). Not too ironically, this is exactly the "slippery slope" many have feared if they accept the environment as a concern and actionable priority for the movement, that is, a slow slide toward liberal views on same-sex marriage and abortion. When Cizik resigned, he stated that while views on same-sex civil unions had changed among a younger generation of evangelicals, it had not among the vast majority of evangelicals, so he would resign because he no longer spoke credibly for all evangelicals.

After he left the NAE, Cizik became an Open Society Fellow, and in a 2010 panel on "Dissent in the Workplace," he described the events leading up to his resignation (Fora.tv 2010).[9] He said that his political views had changed but not necessarily his religious or spiritual views. He said that he saw legal recognition of gay civil unions as a way to protect traditional marriage between a man and a woman. After the NPR interview, he said he wasn't immediately aware of how radical his suggestion had been and that his resignation was requested because of the firestorm it had set off in "the heartland." After more than twenty years with NAE, Cizik founded a new organization in 2010 called the New Evangelical Partnership for Common Good.[10] The group published a book called *The New Evangelical Manifesto* in 2012 with essays from Cizik and Ball.

Another major change was that the group Flourish was formed under Rusty Pritchard and Jim Jewell, one of Creation Care's less visible leaders and the initial public relations representative for ECI. The website statement for Flourish says it was formed with the express interest of making environmental concern *apolitical* for evangelicals (Neff 2009a, 2009b). Katharine Wilkinson's account of Flourish describes it as a split from those profiled in this chapter who led the drafting and signing of ECI. Wilkinson differentiates Flourish as an ethics-based advocacy approach that calls for "broad changes in thought and action," and ECI as an issues-based advocacy approach that focuses primarily on "a specific topic, placing secondary emphasis on effecting a change in values" (119). Wilkinson was unsure whether these divergent approaches would result in conflict or collaboration at the time her book was released in 2012, and it's not clear even now whether these distinctions represent a fissure or not in the movement. Multivocality is generally a hallmark of emergent social movements, particularly as new media make multiple articulations widely available. So while this chapter is being written in light of these changes, it relies on data and interviews gathered during the initial years following the ECI and seeks to unearth the relationships and views that informed the formation of and rising visibility of Creation Care at a crucial time in its development.

Of the several groups I studied, the leaders of Creation Care were the most difficult to establish an ongoing relationship with. Before I went to the first Creation Care conference in Florida in 2008, I had nearly given up on including them in this research project. That conference, however, provided me with an opportunity to set up and record an essential key interview and conduct many interviews informally with those on the leading edge of the movement. It also provided clear evidence of the state of the group's nascence—the conference was small and had the feel of an insurgency set to begin its work through deliberate means and messaging. I had initially thought this group would be the easiest to contact, but I underestimated both their less formalized structure and the general reluctance to speak with a graduate student not affiliated with a Christian college or any of the usual Christian networks. My MIT affiliation intrigued some (generally speaking, it was a surprise that an individual from a strong science school would take an interest in their group), but there was also a substantial amount of suspicion regarding how scientists regard and portray Christians.

What was surprising to me as I began to dig into the context under which Creation Care was being nurtured were the massive structural and institutional changes the evangelical movement and church were undergoing. When I asked Cizik about whether reports of a divide between conservatives and progressives within the movement were legitimate observations, the former VP for the NAE brought to my attention the Pew Forum on Religion and Public Life's surveys among evangelicals. Pew divides evangelicals into traditionalists, centrists, and modernists—distinctions that signal some of the fracturing and change beneath the surface of NAE claims to a voting block of evangelicals over 30 million strong. Pew also clearly demarcates white evangelical Protestants from black evangelical Protestants—a distinction that held in my experience at the Creation Care conference where the crowd was primarily white and middle-class, but Flourish and the "evangelizing" trip I reference in my conclusion do feature African American pastors. Another clear distinction is between mainline Protestants (Lutheran, Episcopal, Anglican, Methodist, etc.) and evangelical Protestants (Baptist, Pentecostal, Evangelical Free, etc.). In terms of these distinctions, then, Creation Care targets evangelicals and primarily white evangelicals who usually haven't been associated with either environmental or social justice issues the way mainline or black evangelical churches have been.[11] But clearly, as the movement continues to evolve, this could change.

As a way of orienting myself to the terrain that Creation Care was working on, I attended Cizik's small traditional church on the outskirts of D.C. and then Joel Hunter's mega-church in Orlando. Cizik's church of less than 100 people was intimate and friendly. Hymn books were used, and the sermon managed to weave news from the Middle East into a message about Christ as the "Prince of Peace." Hunter's, on the other hand, was enormous—I slipped in unnoticed to a large dark balcony where no one sang the songs of the worship band, and Hunter delivered part of a series on Christian obligations to the poor. A soloist sang the Peter Gabriel song "In your eyes" afterward while images of distant impoverished countries and children were projected. The gap between the style, presentation, sermon, and attendees at these Sunday morning services reveals something of the divide between traditionalists and modernists with this segment of evangelicals, as did the response and scuffle over the initial events of ECI (and Cizik's resulting inability to affix his name to the ECI declaration). But upon reading more closely, these incidental events and exposure and even the Pew distinctions only hint at the diverse, deep-seated changes under way that affect the nature of how church is experienced and organized, views on how Christians should be active in civic and political life, what priorities the evangelical movement should be focusing on, and who they should partner with to achieve these goals.

Demographics and technology clearly play supporting roles in some of the shifts. Younger people subscribe to issues such as the environment more closely than the traditional foci on abortion or homosexuality (Banerjee 2008; Cox 2007; Grossman 2007; James 2008; Pew Forum on Religion and Public Life 2007). And the Internet provides a new way for mega-churches to expand or reach their flock via blogs, social media, and webcast. But there is something more essential at work as well—something nearly captured in Joel Hunter's book *Right Wing, Wrong Bird* or Jim Wallis's *God's Politics: Why the Right Gets It Wrong and the Left Doesn't Get It* (2006).

There appears to be a general fatigue in some quarters with the legacy of the so-called culture wars in the 1980s and 1990s and political alignment with the Republican Party as the party of default, replaced instead by a willingness to build coalitions on issues like poverty, AIDS, and the environment (Broder 2008; Little 2005; Roberts 2006a; Salter 2007; Sataline 2008). This is an observation made by many inside and outside the movement, and those three issues are generally lumped together as abundant evidence of widespread coalition building with left- and right-wing political groups. Cizik, in my interview with him, added human trafficking, the civil war in the Sudan, and other pertinent issues.[12] Cizik and Ball were both very clear, however, that Creation Care is coming not from the fringes or progressives of the evangelical movement. Rather, it is coming from and targeted at the traditional conservative heart of the movement.

Despite these indicators, the changes, much like any of a social nature, are hardly even, instant, or unilateral. Polling throughout the period of study up until the present reflects the challenge Creation Care is up against. Among people of faith, white evangelical Protestants are the least likely by a large margin to have been convinced of human-induced climate change. A 2008 survey by the Pew Forum on Religion and Public Life found this group (31 percent) was more likely than the average American (21 percent), and much more likely than mainline Protestants (18 percent) and black Protestants (15 percent) to deny the existence of climate change and anthropogenic causes. And while 47 percent of Americans acknowledged there was "solid evidence" of climate change and human causality, only 34 percent of white evangelicals and 39 percent of black evangelicals agreed. These percentages are lower than that of Republicans in general who are not convinced of the fact of climate change. During this same period, the percentage of Republicans convinced of climate change began to decrease from 62 percent in 2007 to 49 percent in 2008 as compared with 84 percent of Democrats and 75 percent of independents (Pew Research Center for the People and the Press 2008).

Starting Points: Trust and Messengers for Climate Change

My conversation with the pastor and his daughter proved to be a starting point in trying to understand what lies within this disparity. In my interview with Jim Ball, I asked him about this exchange specifically. I told him that a prominent pastor had told me that science couldn't be the basis upon which to convince evangelicals about climate change, and he replied:

Well, in our community, there's obviously been—for over a century—there's been bad blood, shall we say, between the scientific community and the evangelical community. . . . Scientists aren't necessarily the most trusted of messengers in our community, and yet who have been the three main messengers on climate change? Just think about it. Environmental groups. They're distrusted in our community for a variety of reasons. Democratic politicians, distrusted. Scientists, distrusted. So we have this problem—some of us are saying, "This problem is huge, and yet our community is not accepting it because of the people who are talking about it." So how do you get them to accept the message? Well, you need trusted messengers.

This, perhaps more than anything I encountered, gets at the core issues confronting Christians seeking to enroll their churches in action regarding climate change. As Ball put it, the messengers matter whether they speak for the need to act, a moral/ethical basis to do so, or the scientific facts. Scientific facts are not set apart from its institutions, various instantiations and findings, histories, or interactions with evangelical groups.

The declarations and the ECI established a group of credible evangelical leaders who were willing to be the trusted messengers and say to others, "Take this problem seriously." Such other messengers circumvent the problem of having scientists, Democrats, and environmental activists deliver the message about climate change, whether it be the science, policy, or need for personal action. As Ball puts it, they "bless the facts."

We have this strategy of reaching out to evangelical leaders and then eventually they issue a statement saying take this problem seriously for these forward issues, and in effect they *bless the facts*. They allow people in our community to say, "Well, you know, gosh, I don't know about those scientists, but this person I respect does. They made a conclusion that this is a problem and that we as Christians need to address it. So, okay, I'll listen to that." There are still those in our community who are actively opposing. I just forwarded an e-mail from a pastor who got a DVD from one of our allies who's saying that climate change is a serious problem. The pastor writes back and says the science is incorrect. It's a bunch of baloney. My colleague is like: how arrogant can this be that this pastor is saying that he knows science better than the experts. But there's all this distrust in our community.

Ball acknowledged that those actively contesting climate change are not the group that Creation Care is aiming to influence. Rather, their strategy is aimed at people who haven't given the issue any time or attention or had it explained to them as an issue that connects to Christian responsibility and morality.

Ball went further later on in our interview to explain that groundwork must be laid collectively as well. What and how evangelical communities are talking about the issue influences how scientific facts get taken up.

> There's all kinds of barriers for us in getting people to accept this. The three main ones have been the three main messengers. It is not just, of course, about the facts for any of us; it's kind of the social cultural milieu in which you live and exist. And if your friends who you trust are saying you're going to wreck the economy; you're listening to your radio talk shows and they got people in there that say you're going to wreck the economy, and it's the new form of communism, then it's really easy for you to just say I don't have to worry about that. So we've been trying to find people who can kind of burst through that a little bit, and get people to take a second look.

This lack of a straight line between scientific fact production and reception is a crucial step often overlooked by many scholars intent on designing models that address the public understanding of science or science literacy (Irwin and Wynne 2004; Jasanoff 2005; Latour 2004). Public understanding of science models tend to focus on how much or little the public understands the facts and how the public deals with uncertainty or risk, ignoring the multiple contexts within which facts circulate. Ball's point about the importance of the "social cultural milieu" rejects the idea of such models for understanding, but his formulation goes further, touching on notions of framing that have been used to diagnose problems with public engagement of climate change and science.

Framing is a term used colloquially and in scholarly writing to denote what Irving Goffman first observed were the unconsciously adopted cognitive structures that work to govern the perception and representation of events. Framing certainly helps to explain some aspects of how events and issues are perceived, and the media play a role in developing initial frames (D'Angelo 2002; Entman 1993; Gitlin 1980; Scheufele 1999). The process of moving from the lab through the media such that the public becomes engaged is often characterized as one where either the media can be made to see or say things in a certain manner or that their audiences can (Epstein 1996; Gamson 1992; Lakoff 2004; Snow and Benford 1988; Snow et al. 1986). Market research tactics also depend on a certain amount of instrumentalization, and these too have made their way into sociological analyses of climate change attitudes among the public (Leiserowitz, Maibach, and Roser-Renouf 2008). Such analyses can be helpful in drawing conclusions about what media have done and perhaps in how or where to devote advertising dollars as well. But these expansions and adaptations of Goffman's idea of framing often deny a social group's multivocality and/or turn on the idea that the making and reception of meaning can be predetermined and stabilized (Fisher 1997; Steinberg 1999).

What Ball's formulations point to is a much different process by which frames come to be set in motion—and also that frames *are* in motion, always and constantly, being squashed up against other social processes. Part of the challenge faced by all of the groups researched for this book is a confrontation with exactly this open-ended process of how issues come to have meaning and demand engagement owing in part or whole to the code of morality and/or ethics inherent (Latour 2004a, 2004b). In the case of Creation Care, how the fact of climate change comes to matter is a process that Ball is saying revolves around issues of trust and communal reinforcement over and above the weight of facts by themselves. Borrowing from the history of journalism and civic life (Habermas 1962; Terdiman 1990; Warner 1990), it is a process much closer to that in earlier eras of voting and party politics. Schudson (1998) points out that before American voting reforms in the late 1800s that emphasized information as the key to forming an opinion, citizens were drawn to issues, political parties, and voting as a result of their social ties and affiliations. What this book argues is that the social matters—it matters in terms of directing attention, adjudicating debates, knowing who and what one can trust, and what side one wants to be associated with. It also matters, Ball is saying, when it comes to lack of concern, turning off, or *attentional rest*. By attentional rest, I mean to signal Ball's description of those for whom it is easy not to pay attention—even when facts are known—due to a range of factors.

Trust or distrust, then, is the primary issue or terrain upon which much of the movement's potential and credibility lie, and it is established both through mobilizing a vernacular familiar to the movement and through clear identification, standing, and recommendations from and within the group. In other words, part of the problem is science. But politics, a biblical mandate, morality, and guilt-by-association with a liberal agenda are all factors in motivating (or not) evangelicals to make climate change a pressing issue. Yet, as Cizik pointed out to me, if evangelicals take up this issue and recognize it as one of their own, the potential is immense. Consider evangelical activism around other social issues and the potential effect on policy and personal action, as well as party platforms and leadership. These are the kinds of stakes at play when it came to drafting and releasing the ECI.

Examining the Evangelical Climate Initiative

Released in 2006, the ECI is an inherently political and civic document as much as it is a statement of religious beliefs as they relate to climate change and science.[13] It begins by establishing the political presence of evangelicals and its biblically dictated obligation to continue that presence:

> As American evangelical Christian leaders, we recognize both our opportunity and our responsibility to offer a biblically based moral witness that can help shape public policy in the most powerful nation on earth, and therefore contribute to the well-being of the entire world.[14] Whether we will enter the public square and offer our witness there is no longer an open question. We are in that square, and we will not withdraw.

Cizik argues that evangelicals must be near enough to the seat of power to speak truth to it, and this opening salvo demonstrates this belief and the ongoing practice associated with it. The statement also speaks to the underlying philosophy that belief and action are twinned—that knowing something means there is an obligation to act on that knowledge. The ECI goes on to make these strong claims as follows:

Claim 1: Human-induced climate change is real
Claim 2: The consequences of climate change will be significant and will hit the poor the hardest
Claim 3: Christian moral convictions demand our response to the climate change problem
Claim 4: The need to act now is urgent. Governments, businesses, churches, and individuals all have a role to play in addressing climate change—starting now.

There are three key observations to take from these claims. First, at the Creation Care conference, the clearest rationale for acting was based on claim number two. Joel Hunter continually reinforced this message as the host by using the phrase "the least of these" and, at one point, even saying that people will die as a result of lights being left on. This is a distinct part of the work that Creation Care is engaged in—that of nesting environmental concerns within the sets of concerns evangelicals are already engaged with—in this case, poverty, but also increasingly, if Ball and Cizik have their way, the pro-life stance.

There used to be a picture on the EEN site that shows Cizik and Ball marching in a pro-life rally that says: "Stop Mercury Poisoning of the Unborn." In my interviews, both of them mentioned this nesting of an environmental issue within the much more established concerns about abortion. Writing in response to Grist.org's coverage of evangelicals on beliefnet.com, Ball put it this way:

> As increasing numbers of rank-and-file evangelical Christians understand more deeply that reducing pollution is loving your neighbor, as they become more aware of mercury's impact on the unborn, that one in six newborns have potentially harmful levels of mercury in their blood, as evangelicals become more aware that global warming is real and is projected to harm and even kill millions of the world's poorest, whom Jesus Christ identified with himself (Matt. 25:40), they will become more engaged.

It may seem like a leap, but this is exactly the kind of rationale Creation Care engages in when it seeks to convince evangelicals that the environment is already a part of their suite of concerns—they just haven't realized it yet.

Second, for scientific evidence related to claim number one, ECI cites the Intergovernmental Panel on Climate Change (IPCC), noting Houghton as an evangelical Christian who has been involved, as well as the U.S. National Academy of Sciences (NAS) and President Bush's declarations at the 2005 G8 Summit. ECI calls for a plan of action to emerge on the reduction of fossil fuels and tougher national environmental laws, and then it points out specifically the work done by a number of major multinational firms, closing with a call to individuals to help the poor by reducing their carbon impact. Ball says that mentioning business leaders was a specific part of their strategy because evangelicals may not be willing to listen to scientists, but they are likely to pay attention to someone from General Electric.

> They don't understand. The IPCC is what it is. They've been told that it has problems or something and they really haven't thought through:

how should science be used in a policy context and what constitutes enough evidence to say we should take action? They haven't thought through all of that, but when they hear a business leader saying it's a serious problem, the head of General Electric, a business they know, and the VP of Shell or someone like that, that's when we'll start to have people pay a little more attention and be a little more receptive and open to listening to what we have to say.

Economic concerns play a large role in the criticism expressed by opponents to ECI and Creation Care. Knowing that businesses are taking climate concerns seriously negates many of those arguments without having to address them.

Third, in the claims put forward in the ECI, there is a lack of division between policy and personal action, belief and consequences, morality and the demand for response. I would argue that this is distinctly American and evangelical in that evangelicals are active on strongly held beliefs *and* perceived as politically powerful in a system that often caters to and rewards powerful special interest groups, and they are able to have some effect on the system of lawmaking as a result. For example, in a post two years after the 2006 launch of ECI on the *Deep Green Conversation* blog run by EEN, Ball wrote that he and Cizik had been instrumental, along with other members of the National Religious Partnership for the Environment (NRPE), in moving the Lieberman-Warner Climate Security Act toward helping the "poorest of the poor" with climate change adaptation (Ball 2008).

In the meeting with Senator John Warner that Ball describes, it was made clear that the evangelical position was important. Cizik then mentioned an EEN poll that said 84 percent of evangelicals were now in favor of climate legislation. "Rich [Cizik] helped him [Warner] understand that the evangelical community was changing and growing in its concern for the poor and for God's creation." What happened next was nearly unprecedented in terms of religious advocacy in Washington, as Senator Warner allowed them to articulate exactly what they wanted in terms of international adaptation and wrote it into the bill. Ball credits the Holy Spirit for this political win, and he savored especially the fact that ECI had been going strong despite opposition from "some of the most prominent politically conservative Christian leaders."

Certainly, among all groups there is a sense that shifts in public opinion lead to changes in public policy—for example, in chapter 1, Sheila

Watt-Cloutier, former chair of the Inuit Circumpolar Council, said this was why she participated in the work of ICC and media coverage. But it's difficult to track this cause and effect chain that is something like a mantra among activists generally. With evangelicals, there is some reason to believe in a correlative effect. And certainly this effect is what has driven the media and public relations strategy related to the ECI.

When the ECI was released, they went directly to the public with their message both through "earned" media as Ball calls them, or news articles, and through full-page ads in a rather diverse lot of publications: the *New York Times, Roll Call, Christianity Today*. They also ran radio and television ads on Christian and Fox stations in states with key congressional campaigns in 2006. Ball explained that having articles written about them is the best option because it's free and it's viewed differently by people; getting the attention of mainstream press has always been a part of their strategy.

> Evangelicals read secular papers, they watch the news like everybody else. So if you are getting your stuff in the mainstream media, you're also reaching a lot of evangelicals. But we're also interested in changing our community for the purpose of changing policy. So if not only our community, but other audiences like policymakers start to think that evangelicals are becoming concerned about certain things, then they start paying attention. And it's easier to get meetings on the Hill.

This is not a trivial point in terms of media analysis. ECI has not done a lot of work through new media, but articles in mainstream media were targeted at the still large swath of the newspaper-reading public (on or off-line) and even perhaps politicians and their staff. The perception of evangelical interest in this issue clearly began to increase with mainstream media coverage. The blogosphere picked up on these articles as well, providing more scrutiny, debate, and some of what Bruno Latour (2004) has called "instant revisionism." Christian blogs are numerous as are climate blogs, and occasionally they overlap where environmental concerns are being debated and discussed.

Would Jesus Sign the Evangelical Climate Initiative?

To understand ECI more fully, it's worth dialing back to the 2002 campaign for "What Would Jesus Drive?" (WWJD), and beyond that, to the founding of EEN. ECI is not the first declaration of its kind, but it is likely the most controversial. WWJD was undertaken in conjunction with NRPE, an organization that brings together Jewish, Catholic, and mainline Protestants as well as EEN.

EEN was begun in part by the 1990 letter Carl Sagan organized, signed by thirty-two Nobel Laureates, titled "Preserving & Cherishing the Earth: An Appeal for Joint Commitment in Science & Religion." That year also included the passing of the Clean Air Act and an address by Pope John Paul II on the World Day of Peace calling for environmental responsibility. Two years earlier, the IPCC had been formed, and *Time* had called 1989 the Year of the Planet. In 1992, the Earth Summit was held in Rio, and a forum held by the Au Sable Institute and the World Evangelical Fellowship's Unit on Ethics and Society began the discussions that would lead to the founding of EEN (DeWitt 2007b). The following year, EEN released a formal "Evangelical Declaration on the Care of Creation"—a declaration that was even more widely signed than the ECI and that still guides EEN, according to its website. In other words, EEN came out of the same crucible for environmental change and fervor that molded many secular efforts. And it went through the same down cycle that beset many such organizations between the mid-1990s and mid-2000s, when public interest and concern waned. EEN, at that time, was reduced to one staff member, Ball, but that eventually changed in part because of creative approaches like WWJD.

Ball was shocked by the instant and overwhelming attention that WWJD got—on the order of 6,000 of what he calls "earned media" articles, meaning articles written because the campaign was considered "news." Ball explained his benchmark for its success like this.

> I said [before the launch] I'll know we've been successful when there's been a joke on Letterman and somebody talks about it on the Senate floor. Then I'll know that we've kind of penetrated into the public conversation. And I thought that joke on Letterman would happen six months after it was launched, that it would kind of bubble around. We hadn't even publicly launched the campaign, and there were a couple of jokes on Leno.

Ball acknowledges, however, that media attention such as WWJD received in terms of immediacy, longevity, and volume isn't likely to happen again. He was completely caught off guard by the response to WWJD. He said getting listed on AOL, a major player in online news at that time, as one of their top five stories was a primary catalyst that fully ignited the media craze. I got the sense when he was describing it to me that he was still in awe.

"What would Jesus do?" was a common refrain heard in evangelical circles long before EEN chose to torque the last word of the phrase. Ball said, "The basic goal really was to try to have people start to think in our community and elsewhere that transportation is a moral issue and get that conversation started." Christians are certainly not the only ones driving SUVs in the United States, but they do represent a large number of Americans who are open to thinking about their own personal responsibility and moral standing in a certain way. Bringing together environmental knowledge and an awareness of the effect of emissions on climate with the morality that Christ represents starts a very different kind of conversation than one that science alone can start.

The experience with ECI was completely different—though not in terms of media interest. NRPE hired the public relations firm that handled WWJD, and Ball said they took some criticism for the handling of a few events. So when it came to ECI, Ball said they hired an evangelical group, Rooftop Mediaworks, run by Jim Jewell (now the cofounder of Flourish) and based in the suburbs of Atlanta. Ball said people started to find out before ECI launched that it was going ahead, but he wanted "the right kind of coverage." He even held a reporter from the *New York Times* at bay, much to his own incredulity. They had to push the launch twice and tried to convince the reporter not to run with the story, but to no avail. The story ran before their official launch. Ball said that the reason they were trying to have an "appropriate launch" is because they "knew it would be explosive and we wanted to, in effect, kind of catch our opposition off guard." Instead, the opposition in the form of the letter from James Dobson and others forced them to receive media coverage prior to the launch of ECI.

Dobson is part of a large interfaith group and active group of evangelicals who worked to counteract the ECI and other efforts to mobilize evangelicals with regard to climate change. The 1999 Cornwall Declaration on Environmental Stewardship, to which Dobson is a signatory among thousands of others, seeks to put forward "sound theology and sound science" as opposed to "the passion that may energize environmental activism" in order to guide "the decision-making process." *Sound science* has become iconic phrasing in the efforts of skeptics to unseat the veracity of scientific research on climate change.

The Declaration espouses three primary points of disagreement: (1) the tendency to "oppose economic progress in the name of environmental stewardship," (2) the denial of "the possibility of beneficial human management of the earth," and (3) the belief that some environmental concerns "are without foundation or greatly exaggerated," including "destructive man-made global warming, overpopulation, and rampant species loss." The document goes on to argue that "public policies to combat exaggerated risks" can "delay or reverse the economic development necessary to improve not only human life but also human stewardship of the environment." In other words, environmental policy will create barriers for the poor intent on economic development, causing them to suffer longer than they should. It goes on to state a list of general beliefs held by "Jews, Catholics, and Protestants" and a list of aspirations that includes an affirmation of liberty, stewardship, private property, and economic freedom.[15]

Cizik describes what evangelicals must undergo in order to believe in the "reliability of climate change" as a *conversion*. He described to me his own conversion and the importance of "blessing the facts," despite his trust in the science.

> RC: Well, I was a skeptic and a bit of a mugwump in the sense that when I was invited to a 2002 Climate Change Conference at Oxford, I said, "No, don't draw me into this." So I told Jim Ball, who invited me to come. "Look, I will come, but don't expect me to join this debate. I don't really have a fight in this." Heard the science and decided, "Wow! This is compelling stuff," and we're not talking here about just clean water. We're talking about the very fate of the earth. I was totally blown away by the scientific evidence that is to me undebatable, unequivocal, on our human-induced impact on all of these issues from habitat, destruction of species and extinction, pollution, climate change, and so on all these levels, I was just stunned and felt I could hardly keep my mouth shut. But it was John Houghton walking in the gardens at the Palace who said, "Richard, if you believe—you need to tell others," and I said, "Well, that'll cause a furor." And it did, but I survived. (laughter)
>
> CC: Did it matter to you that the scientists were Christians?
>
> RC: Yeah, it helps. It certainly helped because it helped authenticate the truth factor here. It's not that I am suspicious of science; I'm not. I'm part of the younger generation of evangelicals that have no fight intellectually between faith and science. But it helped. And sometimes there has to be some of this hand-holding to assure people of faith that this is true.

Although it was the science that convinced Cizik and provoked an impassioned response that sustained later action, it still came as a message wrapped in a Christian conference with a prominent Christian scientist as his guide. Cizik doesn't describe himself as suspicious of science, and yet it wasn't science that caused him to pay attention to the issue—it was his community of faith, his peers in leadership, and scientists who were evangelical and were involved in agitating the Christian community to take another look at the issue. "We need scientists to explain the 'what,' and we theologians to answer the 'who.' The 'who' is God. The 'what' is what's occurring to Earth which we have been mandated the stewardship for, and so scientists help us to understand what creation is saying about itself and about its Maker, so we need scientists."

To further illustrate how "hand-holding" might work, Cizik told me about a conversation he witnessed involving former vice president Al Gore at a conference in Aspen in 2005.

> I heard Al Gore say to a Southern Baptist, "You mean to tell me that the fact there are 900 peer-reviewed scientific articles confirming human-induced climate change. . . . Do you mean to tell me you need to have someone in the leadership of the church authenticate the reliability of those studies?" And the man he was speaking to said yes, and Al Gore said, "That's just utterly amazing."

Gore's *An Inconvenient Truth* is focused on persuading its audiences based on irrefutable scientific evidence and overwhelming scientific consensus.[16] Yet science, as presented by Gore and by the vast majority of scientists and science publications, lacks the kind of lens that Sir John Houghton used when he responded to Bill Moyers's question for the 2006 episode of *Moyers on America* titled "Is God Green?" The clip that Moyers uses to explain the role of Houghton in Cizik's "conversion" is Houghton saying, "The science we do is God's science. The laws of science we use and we study and we discover, they are God's laws, because they're the way He runs the universe." This, to be sure, is not the usual utterance of a leading scientist either in the UK or in the United States. Indeed, the stumbling block that most are hard pressed to bypass is the problem of origin and evolution.

Evolving Relations with Evolution

The pastor and his daughter made direct reference to evolution by talking about the distrust that goes back a hundred years. Ball referred to the issue of evolution earlier as "all this bad blood" between evangelicals and scientists. Cizik in his interview with *Fast Company* blatantly stated: "Many evangelicals think that because they don't believe in evolution, they have to reject the science of global warming, too." Evolution and the debates over it are more than a stumbling block on the way to believing climate change. Debates over evolution are also a vital chapter in the history of American evangelicals and their role in public life.

The history page on the National Association of Evangelicals (NAE) website begins by talking about the Scopes trial in 1925 where "the resulting loss of evangelical influence in the mainline denominations had led many to believe that conservative Christians had vanished from the scene, never to be heard from again." Susan Harding (1991) points out that even though evangelicals or fundamentalists, as they were called then, "won" the Scopes case (Scopes was found guilty of teaching evolution), they lost the public relations battle—being relegated instead to an anti-modernist "backward" stereotype against the image of scientific rationality and enlightened modernity. NAE states that evangelicals began at that time to build a thriving subculture—one that, scholars note, emerged strongly during the Reagan administration in the 1980s and ever more prominently under George W. Bush's administration (McKenna 2007; Moyers 2006). Yet the chapter NAE leaves out is the revisiting of a Scopes-like drama in 2004 with the federal case of *Kitzmiller v. Dover School District*, where intelligent design—what creationism or a literal interpretation of the Bible's account of creation in the book of Genesis is often called—was firmly rejected by the courts because of its religious roots. Remarkably, despite the way Dover became a media spectacle much like the Scopes trial over seventy-five years earlier, there has been a tremendous cooperation between Creation Care leaders and scientists, and in these instances, it seems that the debates over evolution have largely been set aside in an effort to jointly move masses toward concern about climate change.

In 2007, I went to speak with Harvard scientist James McCarthy about his work with the IPCC and the Arctic Climate Impact Assessment, and instead he began to tell me about his work with evangelicals. I was surprised and mentioned the work of John Houghton, thinking he had been the primary scientist that evangelicals had worked with. McCarthy responded that Houghton had been the chair of IPCC's Working Group I in the 2001 report, and he was the chair of Working Group II. McCarthy was well aware of the work Houghton had done after the conference and even of Cizik's conversion.

In the months prior to my interview with him, McCarthy had been part of a roundtable at the Melhana Plantation in southern Georgia with a group of twenty-eight scientific and evangelical leaders that included Ball, Cizik, DeWitt, and Joel Hunter as well as James Hansen, Rita Colwell, Judith Curry, Eric Chivian, Edward O. Wilson, and James Gustave Speth— in other words, some of the leading and most vocal scientists working on climate change–related research in the United States. Together, they spent three days discussing climate-related issues and "searching for common ground." McCarthy told me that "many of the scientists sitting around the table said it was the most important scientific meeting they have ever attended." I asked him why, and he laughed and said, "I mean in terms of advancing the science. It was Joel Hunter who said to me: 'So, how many people do you speak to in a week? How many are in your classes? And, maybe, if you gave a public lecture, how many people would be there?' He said: 'I speak to 7,000 and *they listen to me*.'" In this formulation, the problem of motivation is reversed. Action isn't driven by facts; rather, action is assumed and the work required is to get the facts trusted. For Hunter, trust revolves around the messenger similar to the way Jim Ball earlier described it.

Many of the scientists involved in the meeting represent centers at Harvard, Yale, NASA, and other major scientific and academic institutions, and they have often been interviewed by major media outlets. They represent an elite stratum of scientists who lead in their field and are also able to focus some of their time and attention on public policy and garnering public attention. But many of these same people have expressed frustration on and off the record about the public's inattention to what their research already shows—that we are in the midst of a global environmental crisis of epic proportions. Some, like Hansen, would likely say the crisis is well under way. So it's little wonder that the thought of igniting an existing social movement with such a message accompanied by a burden for action might indeed make this meeting one of the most important of its kind.

The resulting statement released on January 17, 2007, "An Urgent Call to Action: Scientists and Evangelicals Unite to Protect Creation," speaks convincingly about the shared "moral passion" and "sense of vocation to save the imperiled living world before our damages to it remake it as another kind of planet." It states that the protection of life on earth "requires a new moral awakening to a compelling demand, clearly articulated in scripture and supported by science," and it specifically expresses concern for "the poorest of the poor," who not incidentally also inhabit some of the richest areas of Earth's biodiversity. McCarthy explained that the tone of the meeting reflected this partnership with science and a sense of shared goals. He said: "It was so interesting because we could easily, easily spend our time debating things, but what we ended up saying profoundly was that it doesn't matter whether—it does not matter at all—whether life came into existence in this planet in a millisecond or millions of years. It's at risk."

This is exactly the same argument E. O. Wilson put forth in his 2006 book *The Creation: An Appeal to Save Life on Earth*. Wilson, one of the signatories to the "Urgent Call," wrote the book as a letter to a fictional Southern Baptist minister (Wilson was raised a Southern Baptist) arguing what McCarthy noted was the tone at the meeting. In essence, it doesn't matter how it all began; what matters is what's happening right now.

> Let us see, then, if we can, and you are willing, to meet on the near side of metaphysics in order to deal with the real world we share. I put it this way because you have the power to help solve a great problem about which I care deeply. I hope you have the same concern. I suggest that we set aside our differences in order to save the Creation.

This ability to set aside debates over how the earth began and what observable process governs its continued development is not something many in evangelical circles have been likely or known to copy. But Creation Care proponents are evidently hoping that the trust their movement's members place in their leaders can help bridge the gap.

McCarthy told me that shortly after the meeting in Georgia, he was interviewed by Fox News with Richard Cizik.

> The interview was getting a whole lot of "I don't know about this climate change stuff, and I don't think I'll buy it." The interviewer turned to Cizik and said, "How about you, Reverend Cizik?" and Cizik said, "Yes. I trust these scientists." Cizik said (this phrase they're using over and over again): "You do not honor the Creator by destroying the creation," and the interviewer says, "Well, so how many people you think will listen to you?" And Cizik said, "Thirty million."

The process of speaking and having others pay attention to you, especially when speaking with the authority of Ivy League sanctioned and fully funded peer-reviewed research, is something scientists up until very recently have taken for granted. Yet climate change has forced scientists to fully confront notions of trust, authority, and advocacy, and I would add ethics and morality to the list as well. The issue of trust is bound up in much more than the ability to make people listen to what a scientist has to say (Irwin and Wynne 2004; Jasanoff 2005). This is where epistemology plays a much greater role than is routinely acknowledged by those primarily invested in the science related to climate change.

Andy Crouch, a signatory to ECI and prominent columnist for *Christianity Today*, wrote a response to Wilson for that magazine titled "Letter to a Tenured Professor" (2006). In it, he agrees that Christians do share a deep reverence for Creation, much like Wilson and other scientists, but he also states that scientists have not sought to understand *the basis for that reverence*. Crouch argues that Christians like himself and his wife, a Harvard-trained physicist, see no disconnect between learning from the collective efforts of science and holding Christian beliefs. In a phrase reminiscent of C. S. Lewis, Crouch says:

> I have seriously devoted myself, in the amateur fashion of which I am capable, to acquiring and appreciating the vocabulary, methods, and discoveries of modern science. As a Christian, I see no contradiction in wanting to benefit from the collective human effort to understand a universe I believe to be uniquely suited for human life, designed to reward rational inquiry, and crafted to provoke wonder, reverence, and awe from its smallest scale to its grandest.

But Crouch ends the letter by saying the same treatment has not been forwarded, in this case for Wilson, to Southern Baptists. Implicit in this is a reminder that scientific methods, while useful, are not the only way to either know the world or make one's way in it.

A blog post at the Center for Christian Studies (located at Cornell, but operated independently), following on Crouch's column, similarly argued that the division between science and religion have been overblown (Johnson 2006). But even more pertinent to the arguments made here, the blog post goes on to say that metaphysics and ideology cannot be set aside as Wilson is suggesting. Rather, the beliefs and values of Christians are precisely what calls them to care about the environment and pay attention to any threats regarding its decay and demise. In other words,

Cizik's trust expressed in the work of scientists on climate change not only blesses the facts; it does so with the weight of convictions, moral and spiritual. The facts must be addressed not only because they are trustworthy but because the shared belief system and code of morality established by the Bible requires one to take action when such facts are put forward as trustworthy. Hence the switch to a vernacular—or as McCarthy put it, "the phrase they keep using": "You do not honor the Creator by destroying the creation." The expectations of following biblical mandates is thus intertwined with environmental stewardship.

That way of talking about the environment has been the theme of Calvin DeWitt's work as a scientist and committed evangelical. DeWitt is an older man with graying hair, but he comes across as youthful in part because of his fiery demeanor and wide smile. He was among the most approachable at the Creation Care conference, and he was immediately interested in my research. He said a sociologist had once written about him as a catalyst and connector that brought people and concerns together and made things happen. I got the sense that this conference was a marker of sorts (and one he was immensely pleased with) in his efforts to put environmental concerns before Christians.

Later, when he took the stage at the Creation Care conference, he burst into the old hymn "In the Garden,"[17] got the crowd to sing along, and he described the American environmental icon John Muir as coming out of the Scottish psalter tradition. Despite his training as a biologist and zoologist, and his many decades of teaching at the University of Wisconsin—Madison in environmental studies, DeWitt's talk didn't discuss science or liberalism or politics. His message instead focused on biblical references to the delight of "the garden" and the exhortation to not "destroy its fruitfulness." Creation, he argues in his book *Earth-Wise: A Biblical Response to Environmental Issues* (2007a), is the other way, besides the Bible, that humanity can come to know God. He calls those who see the environment as a tableau for revelation from God "two-book Christians." And in his talk at the conference, he reminded the audience that "Jesus always taught on field trips."

In the last chapter of *Earth-Wise*, DeWitt lists the "stumbling blocks" and "pitfalls" many Christians might encounter when embracing environmental concern (it will lead too close to the New Age movement, pantheism, political correctness, support for abortion, a one-world government, etc.). In a move that separates science from these other concerns, the last one on his list is that "Science is necessarily suspect," which he translates to mean science and atheism are too close together. His response, however, is instructive because it points out that evolution and its debates are often used as a tool by those opposing climate science.

> Promoters of doubt about the findings of climatology and environmental science have become expert in playing on the fears and apprehensions of the public. In so doing they have discovered that linking science with the question of the origins of life and with evolution will cast a pall on all science, regardless of whether it has to do with origins and evolution. The result is an assault on science as a principal way of learning how the world works.

He goes on to defend the "tentativeness" and "integrity" of science and the scientific method, and he states that many influential scientists are Christians.

DeWitt has been thoroughly immersed in Christian thought and study as well as the scientific method of studying the natural world. He has written extensively on the Christian need to deal with environmental issues, and he doesn't stray from the story of Adam and Eve or what's recorded in Genesis. Rather, he uses the Creation story as a tableau to talk about the relationship one should have with God or how one can come to know God. At the same time, he expounds on the principles offered through scientific study such as the hydrologic cycle, carbon cycle, and other forms of energy exchange and ecology. It's in this way that he sidesteps a debate about how life began in favor of a framework he encapsulates as stewardship, appreciation, and awareness. "Our ultimate purpose," DeWitt argues, "is to honor God as Creator in such a way that Christian environmental stewardship is part and parcel of everything we do." It's expressions like this that reflect the message of Creation Care, which makes both environmental concern and the urgency to act inseparable.

Political Alignment

These conversations between scientific and evangelical leaders are still very much conversations among those converted to concern about the environment and the need to act on climate change. In talking about the challenges confronting Creation Care, Ball casts the net much more widely than the evolution debates, noting in particular the widespread perception among many evangelicals that scientists have liberal values and are "part of a liberal culture." DeWitt's long list of "stumbling blocks" reflects a similar base of concern.

The example Ball used was based on a conversation he had at a conservative institution where a person came up to him after his presentation and asked how Ball could trust scientists. The man noted that the American Psychological Association's stance on homosexuality was, in his opinion, ideological and untrustworthy. As a result the man's conclusion—and Ball emphasized that this man is not alone—ends up being: "How can we trust them on climate change when we can't trust them on homosexuality?"[18] In other words, even with window dressing that says otherwise, climate change is still science underneath, with all its attendant difference and anti-biblical tendencies. This emphasis on trust acknowledges the social life of facts—that "scientific" can be perceived as "ideological" regardless of how vigorously its conclusions hew to a prescribed method or discipline. Or, in this case, that some facts can be connected to other facts producing a perception of ideological opposition. The consequence is that for someone like Ball, it means that adopting a scientific fact as a truth worthy of acting upon requires a defense and a biblical exegesis to support it.

At the Creation Care conference in Orlando, this was on full display as speaker after speaker got up to say in various ways: "I am not a liberal," or "I'm here to say you can care about climate change and not be a liberal." Most of the Creation Care–oriented books coming from these speakers begin similarly. Or, as the NAE website put it: "Being concerned about the effect of greenhouse gas emissions on the earth's climate does not make you a liberal kook" (National Association of Evangelicals 2008a).

Part of the work to be done then by those intent on seeing Creation Care succeed is in parsing concern for the environment from what is seen as a liberal agenda or group of concerns. This explains the name change itself from "environment" to "creation" and what Cizik explained to me was a reticence to partner with any secular environmental groups until Creation Care is well established.[19] That it is possible to believe that climate change is a real scientific fact and still be a conservative unallied with other liberal causes such as abortion or homosexuality—two key moral issues for evangelicals active in American policy and politics—is a position that is only slowly being established. And yet, as the pastor and his daughter pointed out at the conference and as Ball affirmed, it is the primary way in which evangelicals will become convinced of the need to address climate change both personally and collectively. Those who deliver the message must be trusted in order to "bless the facts" and to provide the moral, ideological, and epistemological underpinning necessary to make the claim that action is required credible and worth prioritizing.

A secondary and related aspect involves establishing a deep connection with the biblical mandate to "tend the garden," as DeWitt and several pastors at the conference put it, and making that interpretive turn known, available, and trusted by pastors who have either ignored the issue or defaulted to the stance taken by Republican-aligned evangelical leaders, like those opposed to the ECI. It is this step that will likely lend the whole initiative what it needs to be seen as biblical and Christian in the cultural sense—something that Christian individuals can be seen to be involved with and ask others to join in. DeWitt's exegesis makes this aspect abundantly clear, and he is certainly one of several who have made a similar effort (Berry 2000; Bouma-Prediger 2001; Robinson 2007; Robinson and Chatraw 2006; Sleeth 2007). If one considers the way abortion became a "Christian issue" or an issue of morality for Christians to take up, Creation Care is trying to affect the same process for the environment. But the environment involves a much more complicated set of considerations ranging from the status of science, a historical association with the Left, the economy and political positions on its handling, as well as its moral and biblical standing.

This, analytically, is one of the most difficult bundles of considerations to puzzle through, and it speaks to the diversity of the evangelical movement. Within the evangelical movement, there exist groups who identify more strongly with social justice issues like poverty alleviation. Tony Campolo is one such prominent "left-leaning" leader who was a spiritual advisor to Bill Clinton and is generally seen as a Democrat, despite his critiques regarding abortion and same-sex marriage. His 2008 book *Red Letter Christians* spawned something of a submovement of its own and is a challenge to Christians to rise above partisan politics to address the environment and other issues. Yet what is largely at stake for proponents of Creation Care is the twinning of what is biblical and what is politically conservative where "Christian values" stand in metonymically for a much larger swath of concerns that run the gamut from population control and abortion to big government. With the Bible as a rationale for acting, the issue becomes less about opinion or political leaning but about what's right and wrong. Jim Ball put it this way in a post on Beliefnet that was responding to stereotypes he saw Bill Moyers and Grist.com using to describe Christian beliefs.

> The main reason many evangelicals have not been as engaged in caring for God's creation as the Bible calls them to be is because in their

minds "environmentalists" are liberals who hold beliefs (e.g., pantheism) and values (e.g., population control) that can be harmful and lead people astray. Indeed, becoming an environmentalist could lead one to become a full-blown liberal, and thus turn away from conservative Christian values and those who hold them. Some evangelicals are also concerned about what they regard as liberal solutions to environmental problems: big government and oppressive regulations. Because environmentalists are perceived to be liberals, anything tagged as an "environmental" concern must be liberal, too. There is an unfortunate guilt-by-association at play: if something is liberal, then evangelicals should have nothing to do with it. (Ball 2007b)

So it's not just that environmentalism is liberal, but also that if it isn't politically conservative or Republican, then it's not Christian. It is an unfortunate "us versus them" position, particularly for a religion whose tenets also include "spreading God's love" in order to win others as converts. Ball went on to say that this barrier prevents many evangelicals from "exploring the richness of the Bible's message on creation-care," which creates more "ignorance and lack of motivation" to act.

Trying to understand the stakes of political neutrality and an embrace of environmental issues led me to read Joel Hunter's work more closely. Not only is he a key figure within the Creation Care movement who counts DeWitt as one of his mentors, but he's also been spearheading many of the structural and political changes within the evangelical movement. His 2007 book *A New Kind of Conservative* was originally published a year earlier under the title *Right Wing, Wrong Bird: Why the Tactics of the Religious Right Won't Fly with Most Conservative Christians*. What Hunter proposes is a structural change that retains the conservative values of the evangelical church, but not necessarily its political leanings. He seeks a kind of overhaul in thinking about political involvement, and it stems in part from his own experiences when he was positioned to take over one of the pinnacles of the Religious Right.

Hunter first came to media prominence in 2006 for his acceptance and then rejection of the offer to run the Christian Coalition, an organization founded by Pat Robertson and led famously by Ralph Reed in the 1990s, to lobby the federal government on behalf of Christian family values. One of the reasons Hunter stated that he decided to rescind his acceptance of the position was that the board of the organization refused to entertain the possibility of taking the organization in different directions—of tackling issues like climate change, AIDS, and poverty. In *New Kind of Conservative*, Hunter describes the incident in much more generous terms than mainstream media did at the time. *Time* ran a story about the Coalition struggling to remain relevant; the *New York Times* painted Hunter's resignation as one of the Coalition's latest difficulties and quoted Hunter as saying there was a new uprising within evangelical circles not necessarily interested "in the passage of certain laws" (Banerjee 2006). The rebuttal from Christian Coalition representatives was that Hunter had subverted the process by which consensus on agenda changes can proceed in their organization, painting him as something of a loose cannon. In the book, while not necessarily faulting the differences he had with the board of the Christian Coalition, Hunter describes their stance of fomenting "fear and anger" on hot button issues as one that stems from the 1970s. Christians, at that time, still reeling from the tumult of the 1960s, were confronted with enormous cultural shifts such as the Supreme Court decision in 1973 on *Roe v. Wade* and the decision to "subtract" (a significant shift to less aggressive language by Hunter) prayer in schools (Hunter 2008, 21).

Hunter goes on to carefully characterize the resulting para-church organizations like those started by Dobson and Jerry Falwell as helping to fill the vacuum on national moral standards and civic duty. However, he sees the "past success of Christian conservatism" as a block in the maturation of the evangelical movement. The foreword by Ron Sider, president of a para-church organization called Evangelicals for Social Action, makes Hunter's point more clearly by stating that "a powerful evangelical center is emerging that is rapidly transcending the narrow boundaries of the Religious Right" (2008, 13). Sider and other scholars who have published on this emerging shift within evangelical circles point to two major examples of this shift: the 2004 "For the Health of the Nation: An Evangelical Call to Civic Responsibility" adopted unanimously by the board of the National Association of Evangelicals (NAE), and the 2006 Evangelical Climate Initiative (ECI). On both of these documents, Hunter is a signatory and has emerged as a key spokesman.

The vision Hunter puts forward is of connecting on ideals and faith and moving forward on issues in partnership with like minds, regardless of their religious orientation. This ecumenicalism was on full display at the conference when one of the first breakout sessions featured Hunter, a local rabbi, and a local imam, and when a keynote was offered by the Catholic bishop for Florida. When Hunter introduced Bishop Wenski, he said that "new evangelicalism is really old Catholicism," implying that the civic involvement of the Catholic Church was both a goal and a model worth emulating. Wenski, in his riveting address, made it clear where he stands: "We are not endorsing environmentalism. Al Gore is not the fifth evangelist . . . and everything said about climate change is not good science just like everything we hear from the pulpit is not good theology." He went on to explain that the strategy of the church is not to impose on the unwilling. Rather, quoting Pope John Paul, he noted, "The church does not impose, she proposes. We have a proposal to make about what helps or hinders human flourishing." Science, he made clear, serves this proposal for common good rather than the other way around. Wenski's main rationale for working on climate change is something Hunter has written about and what Creation Care espouses—that one of the primary reasons climate change is worth paying attention to is that it threatens the poorest of the poor, or "the least of these" in biblical terms. Hunter came on after Wenski and noted that "God has seen fit to give us problems that no one group can solve on their own," and in his jovial positive way he suggested that teamwork was the appropriate metaphor to move forward on good works.

It is this kind of partnering that Hunter is recommending with any and all who are willing to work on issues that require clear moral prescriptions, putting aside the suspicion one has of those considered the "enemy" (he lists liberals, secularists, "United Nation-alists," etc.). He asks:

> What if there was a way to increase our identity and our intensity for right by associating common causes with "the enemy"? What if *conservative* did not just mean emphases on traditional morality, small government and lower taxes, large military and combat readiness? What if *conservative* also meant doing the right thing in compassion issues like Jesus did: healing the sick, feeding the hungry, appreciating the "lilies" (God's creation), and freeing the oppressed? What if believers were also enthusiastic for the furtherance of science and rigorous training in rational debate? (24)

He is perhaps even more clear-cut later on when he says: "Conservative Christians need to be more ambidextrous rather than just "Right" or "Left" oriented. The Bible is more holistic, more fulfilling to all of life's needs rather than heavy-handed on what is morally right or compassionately left" (78–83 original text, quoted in Deep Green Conversation blog).

Hunter is careful to continually pay homage to those who have established the foundation for civic involvement and Christian positions on fundamental issues like abortion and same-sex issues. And, certainly, Susan Harding (2000) has pointed out that evangelicals possess a kind of bilingualism, moving back and forth between their communities and the larger world. But the ambidextrousness Hunter advocates is meant to question the foundations of alignment between evangelicals and right-wing political advocacy. What Hunter is pushing for is nothing short of radical, and the environment is one of the primary lightning rods for the change that he's proposing.

Conclusion

Creation Care has been forced to confront historical and epistemological concerns that sideline scientific evidence and, at the same time, address heated opposition within their own movement that stems in part from political alignment with right-wing positions on climate change and economic issues. Beginning with its name, Creation Care has had to differentiate itself from the environmental movement for fear of being branded as liberal. As Hunter told the *Orlando Sentinel* when he explained why they use "Creation Care" as opposed to "environmentalism": "We're not tree-huggers, we're God-huggers. . . . . We wanted very much to do this not out of an ideology, not out of a political position, but out of a moral obedience of what it says in the Word" (Carlson 2006). Ball made a similar statement when *Christianity Today* interviewed him during the launch of ECI (Blunt 2006).

For many, science itself is seen as ideologically liberal and therefore not the basis on which evangelicals should act. Hence Hunter's crucial definition of the reason to act being about "moral obedience." This isn't to say that science is completely sidelined, but it is paired with the moral in complex and sometimes paradoxical ways in order to achieve the position of "belief" in climate change as a real problem in need of Christian address. Scientific evidence, in the context of Creation Care, acts as *a partner*, rather than as the sole basis for evidence. Through this lens, as C. S. Lewis put it, biblical knowledge acts in epistemological parity alongside, but also supersedes, scientific (and more general civic) epistemologies in terms of establishing not only what constitutes a valid explanation but what that explanation demands morally and ethically. A biblical mandate must be part of what convinces evangelicals of the need to act, and a part of that work means nesting environmental concerns within existing well-defended mandates regarding, for example, care for the poor and the sanctity of life. Trust to speak on these issues—to "bless the facts"—is established in part through one's position within the church and evangelical movement as well as through identifying as a conservative.

When I spoke with Cizik and James McCarthy, both told me about a planned trip to Alaska later in the summer of 2007. They would be bringing six scientists and six evangelical leaders to Alaska to see firsthand the impact of a changing climate on people and the environment. The six scientists would include McCarthy, Eric Chivian, and others yet to be determined, and the six evangelicals would include three convinced and three unconvinced of the need to act on climate change. Bill Moyers from PBS's *Now* sent a television crew along to film the adventure and conversions. The documentary, called "God and Global Warming," condenses the weeklong trip, which includes the group meeting in the Anchorage airport, on bus trips through Alaska, and trying to convince Bishop Harry Jackson, an African American pastor from the Washington, D.C., area who had joined the group. Jackson is labeled as a skeptic who thinks that most of the calls to action on climate change are "alarmist," and he isn't sure what Christians should be doing about climate change right now.

One of their destinations was the much-chronicled town of Shishmaref, Alaska. Located on a barrier island, Shishmaref is one of several Iñupiat villages in serious peril from a combination of storms, coastal erosion, and permafrost melt. Despite the access to scientists and the hard work the group engages in to convince Jackson, who rapidly becomes a central character in the story, it is the Inuit with whom he connects. When he recognizes their plight and the role of carbon emissions in what they are experiencing, he understands the need to act and to take the message back to his church. He remains somewhat unengaged by the science, but the plight of poverty and relocation are a burden he well understands.

Perhaps this was a factor of documentary storytelling and the need to find a character-driven narrative, and perhaps it was an idiosyncratic individual who bypassed the science in favor of addressing the dire needs of individuals he met and their community. Yet this is exactly the kind of transformation that Creation Care has positioned itself to effect. Jim Ball put it this way in his response to Grist.org:

> Those environmentalists who do not share our faith perspective will have to understand that we evangelicals will have some different reasons for addressing environmental concerns. We may use different lan-

guage, like "creation-care," and we may be more comfortable with labels like "conservationist" rather than "environmentalist." And, frankly, we may seem strange to you at times. But once committed to a cause, we can help make a difference.

What kind, how much, and where that difference is made remains to be seen, but certainly what this chapter raises are the stakes of a coproductionist approach. Ball, those on the trip to Alaska, and many others in this chapter are describing how it is that their group is coming to terms with climate change's form of life. Making climate change a Christian concern involves overcoming a resistance to what's perceived as the dominant meanings, rules, and associations—even the need for a change of name indicates the level of compromise required, and a relationship-building exercise that is not merely based on ideological concerns.

This is the challenge that considering other meanings presents—that scientific epistemological grounds don't automatically trump other modes of apprehending and making sense of either the natural world or the human place in it. Rather, compromise is constituted by way of acknowledging the epistemological concerns and challenges that scientific facts present socially, historically, and ethically. Coproduction as an idiom acknowledges the way scientific knowledge both embeds and is embedded in social practices, identities, norms, institutions, and discourses. Yet the stakes of coproduction necessarily involve contending with other interpretive frameworks, epistemologies, and expertise. The next chapter focuses more specifically on how this is occurring within scientific contexts and within efforts to mobilize diverse publics through articulations of risk and economic calculations.

CHAPTER
FOUR

Negotiating Risk, Expertise, and Near-Advocacy

As consensus has formed around climate change, science experts, particularly in the policy world, have sought ways to explain in pragmatic terms what a future with climate change means in economic terms. When Sir Nicholas Stern released one of the first comprehensive reports of this kind, *The Economics of Climate Change: The Stern Review,* in November 2006, it unleashed a maelstrom of controversy in science, policy, and media circles. Commissioned by British prime minister Tony Blair, Stern's seven-hundred-page report focuses on whether it makes sense or not to pursue a mitigation strategy in the face of a range of scientific climate change predictions. It argues that "strong and early action far outweighs the economic costs of not acting" and that "climate change will affect the basic elements of life for people around the world. . . . Hundreds of millions of people could suffer hunger, water shortages, and coastal flooding as the world warms."

Stern estimates that not acting to avert climate change could cost the equivalent of losing 5 percent global GDP annually—a figure that could rise to 20 percent "if a wide range of risks and impacts is taken into account." Reducing greenhouse gas emissions now, in contrast, could limit those costs to 1 percent annually.

By the time the *Stern Review* was released, the UK government had already chosen to pursue mitigation policies, and some critics saw Stern's work as rubber-stamping that choice. Although the most reported debate was among economists, debate also flared among scientists who saw Stern's report as favoring the higher end of IPCC-sanctioned predictions.[1] This is generally what I expected to hear when MIT held a discussion on the *Stern Review*. The event included a long roster of well-known professors who work on climate science, economics, and policy, including Ron Prinn, Paul Joskow, and Harold Jacoby.[2] When I arrived, the room was already packed with faculty and students. I ended up sitting on the floor, on one of the steps leading from the lecture hall seats, which made taking notes a bit awkward, but I soon became grateful for a seat. Before the event even started, the steps around me also became crowded and the doorway area was jammed with those not able to find a seat on the floor. This was a significant crowd, especially considering it was the week before MIT was formally back in session for the spring semester.

Joskow, an economist, went first and explained the report and the kinds of claims it was making. He asserted, using polling numbers as a reference, that climate change is "an issue of education and ultimately of convincing the people." Prinn, an IPCC author and climate scientist, followed, saying that the *Stern Review* was an example of "how an economist interpreted the scientist." He acknowledged that Stern's group used the same approach as the MIT Joint Program on the Science and Policy of Global Change (which he directs with Jacoby, who's an economist) of "an integrated assessment with some attention to uncertainties." He then began a critique of how and where the *Stern Review* was flawed in terms of conceptual uses of data, overstatement of accuracy, an error in assessing hurricane damage, and its bias toward high-end impacts. John Reilly, also from the Joint Program, followed Prinn and continued the critique, noting in particular that it ignores any of the benefits of climate change. Jacoby was next and took issue with the wide range between 1 and 20 percent of GDP related to costs and impacts, citing specifically the income elasticity of energy use. With each critique, there seemed a crescendo building. The next person, however, changed the tenor of the room completely.

John Parsons, an economist, began by introducing himself as from the Technology and Policy Program at MIT. He said that climate change is a "risk management problem," and he said that Stern was one of the first to come at it from a "risk perspective." He put two slides together that showed the steep increase in global temperature with corresponding risk factors and a Stern image showing dollar values, and said "the whole point of the report is to go from this (the temperature slide) to this (the dollar values)." He said Stern had to make some "heroic assumptions" to make that leap: "If you take the average, you don't get large consequences." He said there are two basic processes for "how to intervene in public policy debates." On one hand, scientists and economists can say what they know, inform discussion as far as "reliable science" allows, "unpack key points to be addressed"—hoping that "the public will debate all imponderables and value judgments." On the other hand, Parsons said, they can "produce a bottom line answer" and go all the way, "no matter how reliable."[3] By "reliable," Parsons is referring to the issue of uncertainty. This is not a matter of what's true or false or of falsification—it's a matter of what can be seen as likely and probable scenarios based on their findings. Scientists and economists must "make best ethical judgments," then "turn the crank and spit out cost and benefit numbers." And Parsons said this is what the *Stern Review* did.

When he finished, Parsons received the only applause that afternoon. The speakers who followed him offered an analysis of public polling and a talk on how the *Stern Review* provides a framework for "understanding a global public good." But during the question-and-answer period, the discussion returned to the points Parsons had made. Joskow was first to speak. He agreed that research needs to be "packaged." He told a story about the first time he had to testify at a Senate hearing, and his mentor at MIT looked at his testimony and said he had to double the numbers by changing how he stated them. It wasn't a matter of changing the result or the actual numbers per se but of *how they were presented*, in order to help others understand the ramifications of that particular research and to have an effect within policy discussions. Joskow said that was a major lesson for him about the policy arena. Others concurred with his experience, citing their own experiences.

Engaging the public and politicians, explaining the science, and transforming scientific data into quantitative and/or economic rationales for policy changes with a range of direct and indirect ramifications—these are the difficult tasks that confront those wishing to see climate change addressed in the political and public arenas. That science must be transformed to compete for attention—that it must, in a sense, be "sold" in a busy marketplace of competing interests, ideas, and priorities is something the *Stern Review* and responses to it, however inadvertently, illustrate. What the response to the report also demonstrates is the way scientists question one another, doubt each other's conclusions, maintain epistemological difference and boundaries between epistemic cultures, and still, despite these varied divergences, can and do choose pathways that allow conclusions to be made into a presentable and usable version for publics and polities. Increasingly, as the MIT panel and many a blogger and now social media responses show, this process of closure and agreeing on a pragmatic and "useful" conclusion is more of an open-ended and visible process than it's ever been.

But herein lies the central challenge that climate change raises for scientists in terms of how scientific findings might be applied and made meaningful for society. In order for climate change to be "sold," for the ramifications related to its range of findings to make sense, for the associated "risks" to be established as meaningful and worthy of major address, science must go beyond the facts and maintain fidelity to them. But what does *fidelity* mean in this instance? Prinn, Jacoby, and Joskow were able to pick apart the assumptions Stern used to arrive at a set of factors for considering risk and, at the same time, appreciate what the *Stern Review* was attempting to do. They employed scientific norms of skepticism and scrutiny and yet also recognized the need to educate and convince the public.

Potential economic outcomes of a future with climate change provide an added layer of meaning—going beyond the scientific facts, employing a differently configured epistemology, and transforming the stakes of climate predictions into economic consequences. Yet there is a beguiling slippage in collapsing sciences, especially and including "the dismal one." Economic valuations attached to scientific predictions provide a kind of ultimate tool for making an issue relevant and actionable in the policy and media circuits of American assemblages and institutions, and educating and mobilizing the public about the stakes related to scientific conclusions. As Parsons pointed out, what the *Stern Review* did was take a range of predictions and make them into "cost and benefit numbers" so that policies can be adopted with a full view of the range of consequences and risks inherent to such adoptions. This configuration, as all-seeing as it may feel and sound, produces an enormous number of questions not just about epistemological difference but also about ethics and the weighing and nature of risk—or, as Parsons put it, "best ethical judgments." So configured, climate change exceeds mere fact, and as a form of life it presents a challenging set of stakes for scientists who must negotiate with other interpretive frameworks, epistemologies, and ethical and moral terrains.

Mike Hulme, a leading climate scientist from the UK, points out in his book *Why We Disagree about Climate Change* (2009) that science is being asked to do a job it can't do—namely, to "justify claims not merely about how the world is . . . but about what is or is not desirable—about how the world *should* be" (74). In his analysis of the *Stern Review* and the controversies and critiques surrounding it, Hulme identifies questions about the discount rate and potential damages and catastrophes related to climate change as "judgments which take us beyond observable or predictable realities and about which science is therefore either silent or deeply uncertain" (127). Hulme argues that such judgments are really about how nature and people are valued, and how responsibilities to "future generations" are viewed—and it's these aspects that inform analyses of risk and subsequent decisions about public policy related to climate change. Yet these distinctions between ethics and facts are not readily apparent in most discussions about climate change, the MIT panel included. Instead, facts are organized in order to present or push ethical questions, much like the hurricane discussions in chapter 2 illustrate.

Emergent forms of life always contain ethical dilemmas, but they also raise questions about epistemology—about how we know what we know, and what constitutes an explanation and valid evidence. Such thinking (and doubting) as Wittgenstein points out presupposes that certainty exists for some things, somewhere, and that judgments rely on a whole learned system of judgments (21e). In other words, how we learn to make "best judgments" and recognize facts as problems is part of an epistemological and collective process. What this brings into view are two sets of interrelated issues—the first is around how accessible the grounds for expert claims (and subsequent processes of closure, consensus, and ethical judgments) should be in a public arena, and the second is around the way scientific norms contribute to and rationalize ethical questions and decisions. What hangs over both of these issues and what lingers in all of these chapters to greater and lesser degrees is the use of "risk management" as a conceptual device and epistemic goal in which to situate and understand the probabilities and uncertainty associated with climate change. Attention to and diagnoses of risk compel science experts concerned about the public good to negotiate with what have traditionally been seen as scientific norms of disinterestedness and objectivity in order to engage in a spectrum of what I am terming "near-advocacy," where knowledge of some facts and implications of these facts compel scientists to speak about the ethics related to their findings.

As the MIT panel response indicates, the importance of convincing the public requires that scientists engage in translation and representation of their findings in a variety of settings including media, policy, and other public forums. And as chapter 2 discusses, juggling professional norms of independence and objectivity with evolving scientific findings for public consumption such that risks are diagnosed, assessed, and addressed has been a challenge for journalists as well. In this chapter I want to attend to how scientists understand their role as advising experts and how they view their relationship with journalists and the role for media in representing their expertise and engaging the public with climate change. First, I want to address more directly the issue of risk and risk assessment, which is of particular relatedness to assessment and use of climate science expertise in public arenas.

Risky Business

The sciences in many ways form an epicenter for articulations of climate change's form of life—a wide range of scientific facts are marshaled in order to diagnose and define the problem *and* point toward probable consequences and possible solutions. STS scholars Clark Miller and Paul Edwards (2001) make the point that turning climate change into "an international political issue" has "involved efforts by scientists to alter the conceptual categories through which people understand and value nature" (6). In particular, they cite the ways in which "climate" used to refer to a particular locale's weather patterns, but now scientists view "climate as a set of integrated, world-scale natural processes linking the earth's atmosphere, oceans, land, and life" (7). The concept of climate as global system allows for the emergence of expertise that can explain events in terms of their relation to planetary scale and offer management schemes that address a global interrelated system. Hence evidence across a diversity of scientific practices has been amalgamated under new institutions, new regimes of fact gathering, and new transnational political bodies that produce reports and offer an assessment of risks. The *Stern Review* is but one example of this kind of work; the ACIA and IPCC are other approaches meant to spur policy actions globally and nationally.

Some have called this work "advisory science," and the people who do it "science experts" as opposed to scientists, though advisory scientists often maintain research programs, too. Their abilities as scientists are often what catapult them into an advisory role, but what they are also likely to hold in common is an interest in how science is understood (or not) by "the public" and utilized in some aspect of governance. They value relevance and an application of scientific findings. In this lies a keen interest in the translation and representation of scientific facts. Simon Shackley and Brian Wynne have argued that "advisory scientists (a hybrid science-policy community in its own right) must negotiate their own credibility not only among the policymakers but also within their own research communities whose work they are representing and translating. These boundary relationships shape the representations of scientific uncertainty" (276).

The work that Parsons praised at the MIT event is exactly this kind of advisory science, and responses to the *Stern Review* at MIT and Yale illustrate the need for Stern to defend the credibility of the report's conclusion in its own research community. Yet who or what is this research community? The *Stern Review* and the MIT panel collapse economic and scientific expertise and epistemologies, eliding the differences in professional norms, rigor, types of evidence, and modeling. Increasingly, however, as many scholars and scientists have pointed out, this is the jumble of evidence, claims, and approaches to climate change in the science, policy, media, and advocacy realms.[4]

While economic valuations and forecasts like the *Stern Review* are meant to provide the ultimate in relevance, they also act to submerge or obliterate other kinds of meaning, morality, and ethics under a monetary figure. Such a move hollows out a form of life for climate change that makes it a risk to be managed, made more legible, or even rational in the face of variability and an uncertain future. There are no right answers to defining climate change or its attendant spectrum of risks, and yet the mirage of more information continues to present itself in the distance as if at some point knowing enough or knowing in the right way will guide society toward action. "Packaging" via economic forecasting or other means is thus continually undone and redone—the illusion of legibility and rationalization masks climate change's inherent variability. Doing something or nothing has serious consequences and catapults risk, and its streams of information and information practitioners—intent as they are on making the uncertain more certain—into the realm of the political, moral, and ethical (Beck 1992).

Uncertainty has been one of the key challenges for journalists reporting on climate change. It's hard to say what climate change is without saying what is directly related to it within the geographic locale of the audience or range of their perceived concerns. Applications of economic indicators to climate change predictions are meant, in large part, to address the problem of reaching and convincing large swaths of the public. But within advisory science and policy arenas, "uncertainty" is regarded quite differently than it is in a journalistic setting. As Shackley and Wynne point out, uncertainty doesn't have the effect one might expect or that many have feared—of undermining the authority of climate science and policy experts. Rather, it acts as a "boundary ordering device" (drawing on Star and Griesmer's concept of a boundary object) to reconcile heterogeneity and achieve understanding and cooperation, in order to better devise plans that will reduce indeterminacy and "optimize" the use of resources and deployment of knowledge. Such ordering requires an adeptness that the MIT panel only hints at and that Shackley and Wynne describe in more detail based on their observations of how climate modelers interfaced with policy regimes in the mid-1990s:

> Advisory scientists must continually ask themselves how industry, environment ministries, energy ministries, and environmentalists will perceive and respond to different possible representations of scientific uncertainty. In a politically contested domain, scientists will be upbraided both for not acknowledging the full range of uncertainties and for being far too cautious by overstating the uncertainties. Advisory scientists have to try to anticipate such reactions and adjust their representations accordingly. (278)

The back and forth negotiation at the MIT panel is only part of the credibility process. The other aspect of it occurs among the bureaucratic, political, and advocacy groups that are subject to—and consequently or subsequently in need of—the kind of insight that science advising might provide.[5]

Risk also operates then as what Shackley and Wynne might term a "boundary-ordering device," by creating a rallying point around which compilation of facts might arrive at a cost benefit analysis that rationalizes both the nature of the problem and the expertise required. Leaving aside the problematic of numbers as somehow objective, the risk framing has led to the use of a key metaphor to explain why cost benefit analysis has become a goal. Parsons explained policies and action required to address climate change as an "insurance policy." Insurance is a kind of everyday vernacular that the wide variation of possibilities associated with climate change fits into nicely. Most of us perceive ourselves as having some risk because statistical reports tell us that a certain number of people are going to be in a car accident, be burglarized, have their house catch fire, or experience a health-related emergency. Insurance offers a way to mitigate this risk and offer "peace of mind" in the face of statistical evidence.[6]

I first heard the concept of insurance applied to climate change when Stephen Schneider deployed it in the workshop with reporters and scientists I detail in chapter 2. He used it as a way to get at *what the scientific findings require* in terms of speaking for them and how they might affect economic and policy decisions. Schneider asked the workshop participants for a show of hands as to who had "fire insurance." Almost everyone raised their hands. Then he asked how many people have ever had a fire. Two raised their hands. He said this was an example of where proof and certainty that something will occur is not required in order to compel us to take action against it by purchasing insurance, and the same rationale should apply when considering policy responses and personal actions regarding climate change. Action should be taken in accordance with, and as a result of, the very existence of risk.

Yet such a formulation still leaves climate change open to debates about the degree, location, scale, and full extent to which insurance is required. If we follow the housing insurance metaphor, insurance requires more and more information—a detailed list of what we have and don't have so that losses can be accounted for and remunerated. Expertise, then, is continually required in order to evaluate how much, what kind, and on what scale insurance is required. "Insurance" also implies that risk can be controlled, managed, and accounted for. Ulrich Beck has described debates about the extent and nature of risk as defining characteristics of a "risk society." Beck argues that the very nature of risks is difficult to account for—that visions of manageability are a mirage in the face of uncertainties that go beyond the ability to know. Risk thus heralds change and calls change into being by demanding that institutions, policies, and people deal with a range of predictions and, at the same time, call forth a range of experts who speak for and about the facts and what they might or might not anticipate.

Scientific facts don't speak for themselves, but how they speak, what's considered legitimate evidence and framing for the evidence, who is considered an expert, why, and when are distinctly cultural aspects of how science wends its way to the public discursive domain. Sheila Jasanoff has examined responses to biotechnology, a similarly contentious complex set of science-based problems with moral and ethical contours, in the UK, Germany, and the United States. Jasanoff terms the current processes by which experts emerge and facts come to have weight and influence as "civic epistemology." Each national context has "a mix of ways in which knowledge is presented, tested, verified, and put to use in public arenas" (258). And she notes that civic epistemologies rely on a concept of "public life, in part, as a proving ground for competing knowledge claims and as a theater for establishing the credibility of state actions."

In contrast to ideas about science literacy and models for the public understanding of science, civic epistemology implies that citizens are implicated as passive *and* active participants in testing expert claims, *and* as producers of (sometimes competing) forms of knowledge. Jasanoff characterizes the United States in particular as having a "contentious" civic epistemology where "truth . . . emerges only from aggressive testing in a competitive forum," liability is a foremost consideration, and it is assumed that citizens can and will test claims. Claims are usually presented in "the language of numbers" in order to make some claims on reliability and objectivity, and credible professional experts must be perceived as free of bias.

Setting the *Stern Review* in historical context, it is hardly the first or only risk assessment to interpret scientific findings in economic terms. Jasanoff sees the quantitative techniques associated with "risk assessment" as emerging in the 1980s in part as a result of U.S. Supreme Court decisions that demanded quantitative rationales for regulatory decisions. She refers to risk assessment as "a highly particular means of framing perceptions, narrowing analysis, erasing uncertainty, and defusing politics," despite the many instances in which it has been shown to be highly normative (266). Numbers are seen as objective and value-free, and as Theodore Porter (1992) has pointed out, they are repeatable and appear "disinterested." But this also has particular ramifications for expertise and public engagement.

It is not by accident that the authority of numbers is linked to a particular form of government, representative democracy. Calculation is one of the most convincing ways by which a democracy can reach an effective decision in cases of potential controversy, while simultaneously avoiding coercion and minimizing the disorderly effects of vigorous public involvement. (28–29)

Numbers act as a resource for closure and removal of an issue from the domain of public debate. Yet while cost-benefit analysis resulting from risk assessment has acted to structure bureaucracies and forms of expertise over time, as Porter details, it certainly has not removed ethical or value judgments.[7] Indeed, the MIT panel revealed as much in the persistent critique of the particulars of how the *Stern Review* arrived at its quantitative evaluation of risk.

Much like journalists who cover climate change, scientists are required to employ a theory of the social, or at the very least, a theory of what role they think experts should play in public and political fora. It's in this space of gauging and heralding risk that stances related to near-advocacy are formed and circulated.[8] Climate change as a set of amalgamated predictions even without dollars attached to them demands an ethical positioning of both fact and expert, but from Merton to Michael Polanyi's concept of a "republic of science," this has generally been seen as outside the purview of scientific norms and practices. In his exploration of many ways in which climate change is deployed and understood, Hulme offers a view of scientific knowledge and scientists as situated and interacting with social and cultural forces, based in part on Jasanoff's concept of coproduction.

Far easier would be for science to remain far away from such trouble in its rarefied and autonomous heartland. But those days for science—certainly climate science—have long gone. Science is clearly called upon to speak in and contribute to public and policy debates about climate change and, as it does so, it struggles to find new institutional forms and processes to shape knowledge into a usable form. By taking on this challenge, science also finds itself subject to new forces which can reshape its knowledge and alter its character. As climate science rubs up against society, the nature of scientific knowledge about climate change is modified. And as specific climate policy responses are proposed, challenged, and negotiated within and between nations, how scientific knowledge is viewed by society also changes. (2009, 99)

What Hulme signals is a transformation of the role of scientist in relation to climate change and to scientific facts, methods, and institutions. It's with this in mind that I want to turn now to what American scientists have encountered in the political arena when scientific knowledge has been tested in the contentious fora Jasanoff references in her concept of civic epistemology.

On a "Swift Boat" to the Senate

From 2003 to 2007, Oklahoma Republican James Inhofe chaired the Senate Committee on Environment and Public Works. During this time, Inhofe convened several hearings on topics related to climate change. His final hearing as chair was in 2006; he remained the ranking minority member, but California Democrat Barbara Boxer became the next chair. The 2006 hearing was titled "Examining Climate Change and the Media," and it featured a mix of four skeptics and nonskeptics, all of whom were trained in geology. On the side of those who supported widespread scientific consensus on climate change were paleoclimatologist and geologist Dan Schrag and historian of science and geologist Naomi Oreskes, both of whom I talked with about this research. The hearing was meant to address what Inhofe considered to be the hysterical, alarmist media coverage of climate change. The hearing minutes run eighty-five pages, including testimonies and discussion (U.S. Senate Committee 2006, 108).

Inhofe began the hearing by accusing media of becoming advocates and abandoning objectivity and reporting on hard science, and he presented a number of articles from various papers as evidence, including one by *New York Times* reporter Andrew Revkin on "the middle." Inhofe's concern was that "poorly conceived policy decisions may result from the media's over-hyped reporting." He was followed by other senators who disagreed and agreed (Barbara Boxer and Frank Lautenberg). Then the four expert witnesses came forward. The first was David Demming from the University of Oklahoma, a geologist and geophysicist. Demming is a known skeptic who works at two conservative think tanks. Dan Schrag was next. He departed from his written statement, going over his allotted time, much to the obvious consternation of Inhofe, who instructed him to wrap it up and "cut it short" throughout the rest of the hearing, even during the question period. Schrag started by saying that the media are covering this issue in a very political era, but he, as an earth scientist, sees things differently. He explained the problem with carbon dioxide and other greenhouse gas emissions and expressed concern about computer modeling. He cast the problem of climate change as an "experiment on the planet" for which we don't know the outcome and suggested that the insurance paradigm was the right one to use when thinking about climate change as a problem.

Schrag was followed by Robert M. Carter, a marine geologist and noted skeptic from Australia. He denounced scientists like James Hansen who use such overly complex computer models that "ordinary scientists" can't understand them. Carter was followed by Naomi Oreskes, who explained her research on scientific consensus. Oreskes first walked through a brief summary of the scientific research (stretching back to John Tyndall in 1859) that led to scientific conclusions about climate change. Then she recapped her well-known 2004 peer-reviewed article in *Science*, which she said she undertook to ascertain just how much disagreement there was about climate change in the scientific community.

During the question period, Inhofe praised Schrag for being a leading scientist and then derided him for his involvement with Al Gore and his appearance with Gore at the Moveon.org event that launched the film *The Day after Tomorrow*. He asked why Schrag was getting involved in "the politics." Schrag responded that he was there to point out the problems with the science in the film, since it was so clearly distorted. Inhofe was surprised by Schrag's answer and said so, noting that his staff neglected to inform him of Schrag's position on the film. He asked Schrag what his criticism of the science in the movie is. Schrag said that the sheer lunacy of an abrupt shift in climate change occurring in a matter of days was his target, as well as the mischaracterization of the thermohaline circulation.

Later on, Inhofe questioned Schrag again and said something about Schrag saying earlier that there was no science behind Al Gore's movie. Schrag corrected him, saying that the movie with no science behind it was actually the fictionalized *Day after Tomorrow*. Inhofe then proceeded down a long and winding path toward a question, along the way noting Richard Lindzen's op-ed in the *Wall Street Journal* and the *Time* magazine cover with the polar bear saying, "Be worried. Be very worried," as well as the predictions in 1975 of a Little Ice Age. He ended with a question about the Kyoto protocol. Schrag responded:

> I am not a fan of Kyoto for a variety of other reasons that we don't have to talk about, but Kyoto was viewed as a first step which would be followed by a series of additional steps that would ultimately reduce emissions by a substantial amount more. So showing that Kyoto by itself would only make a small difference is sort of irrelevant to the

point because ultimately Kyoto was only viewed as a small step. (U.S. Senate Committee 2006, 42)

After the hearings, Inhofe's office released a press release claiming victory, namely, that the hearings had "revealed that 'scare tactics should not drive public policy.'" Despite Oreskes's testimony, Inhofe was quoted as saying that "the so-called scientific consensus does not exist." With regard to Schrag, Inhofe stated:

> I was particularly interested in testimony by Dr. Daniel Schrag of Harvard University, who believes that manmade emissions are driving global warming. Dr. Schrag said the Kyoto Protocol is not the right approach to take and agreed it would have almost no impact on the climate even if all the nations fully complied.

Kyoto as "a first step" getting twisted to "not the right approach . . . [with] almost no impact on the climate" infuriated Schrag. He said that after the hearing, he was "really angry," and he called it a "total waste of time."

Schrag is no stranger to the political sphere. Not only is he a professor at Harvard who runs a lab with varied research and the Center for the Environment, he is also a prodigy who published his first academic paper at the age of fifteen. He went on to study at Yale and double majored in geology and political science. He said he became interested then in science and policy, and later, when he came to Harvard as faculty, he ended up with an office down the hall from John Holdren. Holdren, at the time, was advising President Clinton on new research related to climate change, and he offered to pass along anything Schrag was doing. Schrag said he was excited about this "connection with the real world" and the return to his interest in science and policy. Unlike Holdren, Schrag has maintained a vigorous scientific research agenda, publishing eight to ten papers a year. But he remains interested in the policy side of things. Following this hearing, however, his frustration began to outweigh his interest.

Unlike a scientist without the means to talk back, Schrag took his anger and molded it into a searing op-ed for the *Boston Globe* on December 17, 2006, titled "On a Swift Boat to a Warmer World." He said he came to the hearings as a "climate scientist" and an "optimist"—an optimist who believes that "we can fix the climate change problem." He said he knew Inhofe was a skeptic, but he hoped to educate the other lawmakers. He then watched "in horror" as the two skeptic witnesses, associated with industry-funded think tanks, "spouted outrageous claims intended to deceive and distort." It was unfortunate, he added, that "the format does not allow for direct debate." He concluded that while some senators like Boxer had tried to defend the scientific community, "no one stood up and called the hearing what it was: a gathering of liars and charlatans, sponsored by those industries who want to protect their profits." The press release's mischaracterization of his comments only added insult to injury. And it was later, he added, that he found out that Inhofe's communications director, Mark Morano, was behind the swift boat veterans' attack ads against John Kerry when he ran for president in 2004—hence the op-ed title.

It is somewhat paradoxical that Schrag testified alongside Oreskes. Oreskes's research on scientific consensus, in some sense, provided the impetus for that term to become so widespread. It has been criticized by some nonskeptics who feel that science must retain its skepticism— its ability to question dominant examples. Others have pointed out that there is still plenty of debate on the details, making the notion of consensus less unified than the term ordinarily would suggest. However, the utility of *consensus* is clear in the phrasing Oreskes used at the hearing: "Scientists, my study showed, are still arguing about the details, but the overall picture is clear. There is a consensus among both the leaders of climate science and the rank and file of active climate researchers." In other words, "scientific consensus" exists so that scientists don't have to undergo what Schrag did. It is a working term, a slogan word in Ludwig Fleck's terminology, that exists to support the claims and presentations of scientific spokespeople. It exists so that expertise can be redirected from questions of veracity to those related to solutions and policy. Yet these questions are much more difficult to untangle than merely setting a basic tenet of scientific fact straight might entail. In fact, for skeptics, consensus leads right back to the need for more information—to more fully elaborate the problem, the players, and the assemblages that make consensus possible and reliable.

Consensus in some respects has had the opposite effect of clarifying the framework in which to understand climate change. Oreskes pointed out in the question period that she has been the subject of repeated attacks by skeptics:

Since my paper was published in *Science* magazine in 2004, I have received hate e-mail. I have received threatening phone calls. I have been threatened with lawsuits by people who deny the scientific evidence of climate change. So there has been enormous pressure on academics not to speak up on this issue, and it is not just a matter of government science. It goes across the board. (U.S. Senate Committee 2006, 35)

"Packaging," then, as the MIT forum described it, is not enough to intervene in policy or media related to climate change the way it once might have been. Scientists who choose to become science experts like Schrag or Oreskes must advocate for their research, debate its veracity, and endure attacks.[9] They must engage in forms of near-advocacy in order to articulate science as properly open on some counts and closed on others—demarcating which risks are known and which aren't, for a heterogeneous public sphere where forms of life proliferate, compete, debate, and sometimes misconstrue events and claims in order to claim victory for one side or the other.

IPCC: Doing Harm to Science?

In May 2007, on the eve of the release of the fourth assessment report, *Der Spiegel* published an article by Uwe Buse that looked at the way the IPCC was operating in the media. Buse characterized the IPCC's advocacy as "emotionalized" and "hysterical," citing noted skeptic, former IPCC author, and MIT professor Richard Lindzen's description. Emotionalized was the word IPCC's head Rajendra Pachauri had used to describe *An Inconvenient Truth*, so Buse was in effect putting the IPCC in the same group as Al Gore, a politically identifiable advocate.

Buse, while acknowledging Lindzen on his characterization of the IPCC, was not skeptical about climate change. Instead, he posed the question "Is activism trumping science?" Buse felt that the IPCC, instead of acting like an expert advice-giving body like it was designed to be, had become more like a political pressure group aligned with like-minded politicians, pushing for changes in greenhouse gas emissions. He quotes Pachauri and well-known climate scientist Stefan Rahmstorf (a contributing author/blogger to Realclimate.org, a blog coauthored by a group of scientists, including Michael Mann) as saying they see climate change as an "existential issue," and they don't want to be asked why they "didn't do anything" by those dealing with the fallout from climate change a half-century or more from now. As a way to understand such scientists, Buse interviewed Peter Weingart, a sociologist of science from Universität Bielefeld in Germany. Buse paraphrases Weingart as saying, "Scientists usually learn only to reflect on the results of their work, not on their role within the social decision-making process. As a result, they join forces with politicians who share their views. And in this way they do harm to science." It's unclear from this quote where and when Weingart's views begin and end and whether he agrees with Buse's characterization of the IPCC as indeed being active and advocating over and beyond their role as objective scientists.

Scientists being unaware of "their role" in "the social decision-making process" goes against the idea of activism. For certainly, to be politically active on any scale, one must be keenly aware of one's voice and the power inherent. Curiously, the position struck by the reporter for *Der Spiegel* also removes agency—as if scientists should stay stuck as cogs in the on-going big wheels of Mertonian norms, providing expertise without opinion, valuation, recommendation, and/or attention to impacts. Beginning with Fleck and Thomas Kuhn, history of science and STS scholars have continuously demonstrated how the social continually intervenes in the production of scientific knowledge, despite best efforts and/or pretensions otherwise. Scientists are constrained and act within institutional norms, to be sure, but they are in constant negotiation with them as well.

What presents an additional conundrum is the global nature of climate expertise and media coverage. *Der Spiegel*'s criticism of the IPCC and its concern that IPCC activism was "harming" science may or may not be related to the German context and expectations about how scientific expertise should behave (Jasanoff 2005). But what happens when civic epistemologies get translated and circulate their criticism and evaluations of evidence much further? Similarly, the *Stern Review* was produced for the British government within the context of British public opinion and education, and as the *Economist* pointed out, the 1 percent GDP metric is routinely used to explain why action must be taken now. This metric was vociferously debated in the United States exemplifying the multinational nature of expert networks active on climate issues. Moreover, in Fleck's terms, what is a "slogan word" or even a form of life in one country can be translated and redeployed differently in another country or institutional context, creating new frictions and accessing different symbolic power and spokespeople. Yet, despite these global channels of media coverage and expert debate and/or collaboration, scientists still must contend with their national contexts in terms of the ramifications of their utterances, the esteem of their colleagues, and the stability of their funding. The inter- or multinational rapidly becomes national and even local, particularly when dealing with those opposed to action on and/or the veracity of climate change.

I had just read the article in *Der Spiegel* when I went to interview James McCarthy at Harvard University. McCarthy is the Alexander Agassiz Professor of Biological Oceanography. He holds appointments in the Department of Organismic and Evolutionary Biology and the Department of Earth and Planetary Sciences at Harvard. McCarthy has also served in many high-ranking positions on national and international scientific committees. His involvement with climate goes back to the mid-1980s when he was chair of the International Geosphere-Biosphere Program that was part of the International Council of Scientific Unions. He has worked with the IPCC for the past two decades but most prominently as the chair of Working Group II for the 2001 Third Assessment Report. Working Group II assesses impacts and vulnerabilities related to climate change. He was also a lead author on the Arctic Climate Impact Assessment Report that followed the 2001 IPCC reports. When I spoke to him, he was president-elect of the American Association for the Advancement of Science (AAAS) and still conducting research in his lab at Harvard.

I asked McCarthy where he saw the dividing line between science and politics, particularly in relation to the IPCC. He said that climate science necessarily attracts those who have a sense that their work is important to society and those who want to be involved in sharing their research more broadly.

> Well, you know, there are some scientists that want nothing to do with anything like this: "Leave me alone. I just want to stay in my quiet laboratory." But I would say that's a relatively small fraction of people who are working on subjects in this area [climate]. Now that may sound like a hair-splitting distinction, but it's my own personal view that people are working on subjects related to this area because they know there is some importance to that knowledge that is quite different from studying some highly specific phylogenetic analysis of an obscure group of spiders that lives only in Madagascar or whatever and spending your whole life on that, which people do. I think a conscious decision to work in an area like this [climate] is not independent from a sense that this is important knowledge for society. I personally believe that if our science is supported by the federal government, we have a responsibility to share this information.

McCarthy said he wrote something along these lines for his election to the AAAS. He sees the public appreciation of the importance of science as crucial to supporting the ongoing funding system. And he says that while scientists should be doing what they love, the onus is on scientists to allow their research findings "to be used in addressing societal problems when society asks us to do so."

He explained that with the IPCC, governments nominate authors for working groups. When he was chair of Working Group II, he said he had over 1,100 CVs to sort through, and he ended up with eighty or so applicants who he thought could make strong contributions. Those who were chosen made an "investment that would never reward the way a scientific paper or next proposal" would—they spent weekends, nights, and generally worked at a "grueling" pace. He said he felt that "what the IPCC does is advocate for the very best science that we put into the decision-making policy arena." The process then of producing consensus through the IPCC is one that is inherently political, based on government selection and nomination processes in addition to being about the diligence of both contributing authors and the scientist–policy experts who choose from the government-limited pool (Lahsen 2010).

In terms of media, McCarthy said reporters sometimes get it wrong, and sometimes scientists get in a tough position when they go beyond the science due to their own desire to have research become news or as a result of wanting to answer reporters' questions. His own position is that "if I don't think a reporter is going to get something right, then I will not want to interact with them." Still, he said, he can't think of a single instance where he thinks science has been "sullied" by any of this.

The article in *Der Spiegel* and McCarthy's response to it bring to the fore some of the key issues for scientists working on climate change–related research. First, McCarthy claims that scientists drawn to work on climate issues are, for the most part, already keenly aware of their responsibilities toward society and the import and funding of their research. Second, this raises questions about advocacy—about where scientists may choose to channel their energy, and whether such work has an effect on their ability to dispense expertise as required and requested by society. And, last, it raises issues about media involvement and the responses both from the wider public and other scientists when their colleagues choose to advocate for anything, even if what they're advocating is, as McCarthy puts it, "for the very best science" to make it to policymakers.

John Holdren has been a prominent figure in science policy for several decades. When I spoke with him, he was a professor of environmental policy at Harvard's Kennedy School and the outgoing president of the AAAS, as well as the director of the Woods Hole Research Center. He agreed to answer some of my questions by e-mail as he was traveling at the time of my request in mid-2008. Several months after we exchanged e-mails, he accepted the position of science advisor to President Obama. His appointment to Obama's administration was roundly seen as a victory for those advocating action on climate change. Not only has he been outspoken on the issue but his work at the AAAS had included crafting a statement on climate change. Many conservative critics were unhappy because of what they saw as his alarmism, but more specifically, much of their criticism was focused on earlier work he had done with Paul Ehrlich, including writing a 1977 book suggesting highly controversial policies (abortion, sterilization) to deal with overpopulation issues (Eilperin and Achenbach 2008). In between population and climate change, Holdren was also heavily involved in nuclear issues, and when the Nobel Peace Prize was awarded to the Pugwash Conferences on Science and World Affairs in 1995, Holdren gave the lecture. It was titled "Arms Limitation and Peace Building in the Post–Cold War World."

Holdren became an engineer (aeronautics and astronautics) at MIT, and later did his PhD in theoretical plasma physics at Stanford. He has taught and practiced in both areas. I asked him whether he saw himself as a scientist, policy advisor, advocate, or some combination thereof. He noted his training, and then wrote:

> A major preoccupation of mine since my graduate student days has been trying to understand the implications of what we know about science and technology (in general and in specific domains) for crafting solutions to major societal problems that sit at the intersection of S&T [science and technology] with the human condition, notably energy, environment, development, population, and nuclear weapons.

He said it was essential that those with a deep understanding of S&T participate because if they are "disqualified" from participating in "crafting and promoting sensible policies on the problems that cannot be understood, never mind solved, without a deep understanding of the relevant S&T, we are heading for even deeper trouble than we're already in." He traced his own involvement with the climate issue back to a book he wrote (with a journalist) on energy in 1971 where he said the "impacts on climate were likely to be the ultimate constraint on energy use." Not incidentally, that book, *Energy: A Crisis in Power,* was published by the Sierra Club and has been referred to as a work about the concept of "peak oil" (Bailey 2009; Holdren and Herrera 1971).[10] Such conclusions would be seen as a work of advocacy by some, and certainly by most as taking a position on what scientific research means for society and what steps society should take to address this.

Holdren agreed with McCarthy that climate change, as well as other issues Holdren has worked on like nuclear energy and nuclear weapons, attract "those interested in the S&T/society interface and interested in the communication and public education challenges that go with that interface." But he went further, saying there were two subgroups of scientists guided by "different philosophical/ideological/political orientations"— those who see the technologies as posing "big risks" that need to be "properly guided and regulated," and those who feel the bigger danger is "losing the potential benefits" through too much regulation and guidance. He said that many of the same people who work on climate change had gotten involved with nuclear weapons and nuclear energy. On all three issues, he said, "the public stakes are high," and the complexity and interdisciplinarity of the science and technology involved require most laypeople to have to trust the experts on the S&T dimensions. As well, the point about there being two subgroups is not inconsequential—on all three issues, he said, "ideology and political agendas strongly affect many people's inclinations about what should be done (generating many complications around the intrusion of ideology and political agendas into the interpretation, communication, and public understanding of the relevant S&T)."

Holdren comes down firmly on the side of action, regulation, and "guidance," even if that means forced sterilization to deal with overpopulation—a recommendation that he made in the 1970s. It's crucial to point out, then, that Holdren doesn't set one side up as true and factual and the other as false and ideological. Rather, he concludes that "ideology and political agendas" have an effect, "generating many complications" about how facts get communicated to the public—paving the way, possibly, for such a distinction in hindsight, or at least leaving it open for interpretation. As the opening responses at MIT to the *Stern Review* and the article on IPCC activism similarly point out, there are significant issues with how expert opinion and scientific evidence, conclusions, and predictions are presented to the public. Taking this together, and following Foucault's ideas about power and knowledge, it becomes possible to generate much broader questions: What position must a scientist occupy in order to speak and be heard? How are regimes of power and truth reinforced by discursive strategies? What kind of multivocality is possible?

Jasanoff has pointed out that such expertise corresponds to institutional imperatives specific to nation-state contexts and their inherent civic epistemologies ("the criteria by which members of that society systematically evaluate the validity of public knowledge"). In this case, those interested in climate have, to varying degrees, come of age in a context informed by scientists active on public policy issues. And in McCarthy's and Holdren's case, they also feel compelled to do so out of an ethical obligation to address difficult societal problems and/or because their special knowledge allows them both the purchase and obligation to help find solutions. Holdren cited a long line of connectedness between those active on nuclear and climate issues, and McCarthy as well as others I interviewed cited groups like the Union of Concerned Scientists as exemplary forebears and contemporaries. The importance both agreed was that all activism and/or advocacy should stay "consistent" with the "relevant science."

Expertise is a key bridge to policy, media, and advocacy. It's where scientists act as translators, encountering and generating friction. These scientists may advocate for a particular position or set of actions related to what their scientific research has revealed, but this is not advocacy in the strictest or most colloquial sense, and it exists on a wide spectrum of actions. Instead, I'm terming it "near-advocacy" to illustrate the ways in which labeling it as advocacy elides the slippery tasks undertaken of being expert, citizen, *and* scientist, and making science relevant in the policy and media arenas.

Part of what has forced this near-advocacy is the fact that climate change is seen by many as a partisan issue, especially in the United States, where legislation and policy have yet to reflect scientific consensus on the issue. The hearings conducted by Inhofe are one example. This partisan divide informs how and what expertise is sought. Journalists have usually made a strong distinction between scientists based on their independence from advocacy and government entities, but they are increasingly extending strong distinctions to this other element of near-advocacy as well. To put it bluntly, it matters whether or not a scientist is labeled as a skeptic. But stating that there is consensus does not tell us upon what grounds experts have reached their conclusions nor does being in consensus elucidate epistemic approach or difference. This is in part what is at issue in discussions in chapter 2 about "the middle." What it means to be in consensus is not clear for scientists who hold to the ideal that academic scientists should be seen as arbiters who should, as *Der Spiegel* implies, conform to standards of unbiased objectivity and expert, politically uninvolved advice.

Hurricanes and Climate Change

MIT atmospheric scientist Kerry Emanuel subscribes much more closely to the ideal of objective, politically uninvolved advice, and yet his elevation by media and advocacy groups has made that somewhat more complicated. As I note in chapter 2, Emanuel became extremely well known by media and the public as a result of a *Nature* paper published shortly before Hurricane Katrina. It became the basis for many news stories that cited the increase in intensity (which was a correct characterization at that time) and frequency (which was incorrect) of hurricanes in a warmer world. Emanuel was later lauded by *Time* magazine as one of the hundred most influential people in 2006.

In speaking with Emanuel in 2007 and 2008, I got the sense that while he doesn't seek public attention for his research, he's certainly not averse to public engagement. Perhaps this is the lot for scientists who work on hurricanes. During my first interview with him, his audible answering machine picked up several times and reporters left messages, wanting to know his response to this season's hurricane predictions. Emanuel also wrote *Divine Wind* in 2005, a beautifully illustrated book that explains the history and science of hurricanes for general audiences. Yet he draws the line when it comes to advocating based on his research findings. When I asked him about scientists and advocacy, he said that he considered science to be "anti-advocacy" and that it should be available to any group that wants conclusions obtained through science.

> I think the best thing that we [scientists] can do is inform. That's where we're powerful. The problem comes when we become advocates, and that gets to be dubious because it compromises both impartiality and, just as importantly, the perception of it. It makes us close our eyes to new evidence that might come along that goes in the other direction. So we have to be very, very careful to suppress the desire to become ad-

vocates, I think. I don't like the notion of a scientist advocate. I might say I'm very concerned about global warming, and I think we should do something about it. But when it comes to advocating a particular plan, if I'm not the kind of scientist who studies policy and its effects on society, I would be stepping outside my expertise. And I don't think that's wise.

Emanuel then is on the other end of the spectrum from Holdren, certainly. Whereas Holdren sees himself as a vital contributor to policy, Emanuel is more likely to leave policymaking to others who might rely on his advice and research findings. Yet in considering Emanuel's involvement in public spheres, it is difficult to establish where and what is termed advocacy— hence the term *near-advocacy*.

Emanuel underwent the first test he outlines—of having his eyes opened to new evidence—shortly after I talked with him. In April 2008, he published an article in the *Bulletin of the American Meteorological Society* based on new modeling techniques that threw into question what the link between climate change and hurricanes might be (Berger 2008; Emanuel, Sundararajan, and Williams 2008). He summed it up this way to Andrew Revkin of the *New York Times*:

> The models are telling us something quite different from what nature seems to be telling us. There are various interpretations possible, for example: (a) the big increase in hurricane power over the past 30 years or so may not have much to do with global warming, or (b) the models are simply not faithfully reproducing what nature is doing. Hard to know which to believe yet. (Revkin 2008a)

This kind of uncertainty should make it difficult to use hurricanes as an icon for climate change, and yet Al Gore, journalists, and environmental organizations have done exactly that. The poster for *An Inconvenient Truth* shows the eye of a hurricane bleeding into smokestacks emitting smoke, clearly making the visual connection between the emission of greenhouse gases and hurricanes.

It was Hurricane Katrina that allowed many to point to climate change as a real and present danger, capable of inflicting massive loss of life and infrastructure, thanks in large part to Emanuel's paper. Investigative reporting and analysis later revealed that the state of the levees was a major reason for much of the damage. Emanuel responded this way when I asked him about the weight assigned to Katrina.

We've been worried about New Orleans for decades. Katrina wasn't that extraordinary as a hurricane, as a meteorological event. It was extraordinary as a social phenomenon. You can't pin Katrina on global warming because, as I said, its likelihood might be a bit higher because of global warming, and people jumped on that as proof. That is not correct, and yet at the same time, it has to be confessed they're taking the global problem more seriously now. So they arrived at kind of the right answer through the wrong line of reasoning (laughter), and nobody tried to make them do it that way. I mean, nobody in my community claimed—I don't think—that Katrina was a signature of global warming. The press did sometimes, or they interpreted my work. It's a quirk of timing with that *Nature* paper coming out the same month that Katrina happened. It's peculiar the way things work.

Trying to situate one's research among the vast messaging related to the evolving form of life that is climate change is, as Emanuel signals here, a nearly impossible task. Framing when attempted proactively not only instrumentalizes the concept of cognitive models (Goffman 1974) but it also ignores the way messages, frames, claims, and forms of life are continually morphing and ricocheting.

While the "social phenomenon" of Katrina may have happened to Emanuel, he has not shied away from engaging with media and public debate, nor is he averse to some kind of address of climate change. That's not to say he isn't annoyed by how he is sometimes covered. The day I interviewed him, I asked him about a recent quote in the *New York Times* that had him criticizing Al Gore. The reporter had left off a key phrase that torqued Emanuel's quote from "I thought he [Al Gore in his film] overstated the role of carbon dioxide in the glacier cycles" to "Gore overstated carbon dioxide"—a significant change. Emanuel told me: "That's an important distinction to make. I don't think he's overstating it for the future; he overstated its role in those particular variations." He said this kind of de- or recontextualizing of his quotes happens all the time. Emanuel still responds regularly to media requests, but he also wrote a long essay in the *Boston Review* in 2008 that clarified his own stance, his thoughts on the IPCC, and media coverage of climate change. It was later turned into a book.

The *Boston Review* essay makes it clear that Emanuel holds opinions about what's been going on with climate change both inside and outside science. He applauds the work of the IPCC in its assessment of scientific evidence and explains its work and role clearly for a lay public. The essay stops short of suggesting specific policy solutions. It's titled "What We Know about Global Warming," but it does end by delineating the partisan atmosphere in the United States, theorizing why it might be difficult to achieve goals related to climate change.

When I spoke to him, our conversation was much more wide-ranging. He talked about the situation at National Oceanic and Atmospheric Administration (NOAA) and the Hurricane Center. He felt that the Hurricane Center had been turned into a political tool and that the NOAA directives restricting their scientists' ability to speak to media were draconian. Chris Mooney, in his book *Storm World* (2007), profiles Emanuel as well as other hurricane scientists, bringing into sharp relief the differences between empiricists like William Gray and modelers like Kerry Emanuel. Mooney opens the book by describing a scene at the American Geophysical Union (AGU) where Emanuel interrupts his talk to castigate NOAA for restricting its scientists from speaking to the media, and he receives a spontaneous applause from the audience—an audience, it's worth noting, primarily composed of scientists.

Emanuel holds extremely strong views on coastal development. He advises insurance companies, sits on boards, and is well versed in the way the insurance industry works. He's disgusted by the way Massachusetts has handled insuring property along the coast and says that insurance rates include a subsidy for those generally very wealthy people who choose to live on the beach. Insurance companies have roundly pulled out of the coastal market, refusing to insure for reasons clearly obvious to anyone who studies hurricanes. Massachusetts was forced to step into the gap and now provides most of the coastal insurance in the state—at an affordable price, thanks to subsidies from other regions of the state. He was dismayed that journalists hadn't picked up on this story, one that had far-reaching effects in terms of policy and potential losses.

I began asking about this because I was intrigued when he and nine other leading hurricane scientists released a statement on July 25, 2006. It begins:

> As the Atlantic hurricane season gets under way, the possible influence of climate change on hurricane activity is receiving renewed attention. While the debate on this issue is of considerable scientific and societal interest and concern, it should in no event detract from the main hurricane problem facing the United States: the ever-growing concentration of population and wealth in vulnerable coastal regions.

This statement ends by calling for government and industry to "undertake a comprehensive evaluation of building practices and insurance, land use, and disaster relief policies that currently serve to promote an ever-increasing vulnerability to hurricanes." Again, it stops short of calling for specific actions or policies, but it does use science and their positions as leading scientists to call for change. Emanuel said there was no response to the statement.

Is such a statement advocacy or near-advocacy? Is it something akin to what *Der Spiegel* accuses the IPCC of doing, or has the IPCC somehow gone further? Is the IPCC advocacy work different because of the many environmental organizations that support the cause they're associated with and the many that don't? To put it in climate change terms, there aren't any analogues to Al Gore, Greenpeace, Sierra, or WWF looking to rein in and reassess coastal development, nor are there any Cato Institutes or Competitive Enterprise Institutes that call for unchecked coastal development. The lack of response either from policy or advocacy groups makes it difficult for reporters to cover such a statement, to gauge its impact and prominence in shaping either policy or pushes for policy. The blogosphere may yet prove an alternative to mainstream reporting in this respect. It provides a forum and avenue of response for tracking, charting, and intervening in and connecting to many forms of life.

The Problem with "Advocacy"

Given his deep research into hurricanes, I asked science journalist Chris Mooney what he thought about the statement, Emanuel's involvement with insurance companies, and advocacy in general. He said he felt that the scientists who made the statement "didn't have a choice." They couldn't pass that task off to anyone else because "they were the newsmakers." He saw the experience that Emanuel and others who were part of *Storm World* have undergone as "transformative" and as an "uncomfortable shift" into the role of newsmaker. He said he wasn't surprised by the ties to insurance companies, and he saw this as another important role for scientists because of their access to "important" information. "I want our scientists to be providing that kind of information. I don't want them to, in some way, have their work skewed by commercial ties, but I don't have any reason to think that's true. This information is too important. I wouldn't expect them to remain completely unattached from the people who want to use it." With regard to the IPCC, Mooney was similarly cagey about calling them "advocates." Instead, he pointed to the long tradition, like Holdren did, of scientists getting involved on issues—"of taking advocacy positions."

> Do you know how many issues Albert Einstein was an advocate on? There's a very long tradition of the scientific community engaging in advocacy positions on arms control, for example. That in some ways was more political than just saying we've got to raise attention about the fact that the planet is changing. None of them are saying what the policy answer is. It's a tricky interface, but I would be unhappy if scientists never said anything. I think it would be inconsistent with the tradition of what scientists have always done. A lot of the greats in science have had their moments by taking a stand. They have taken some wrong ones, too.

In the end, then, Mooney, who also wrote *The Republican War on Science* and is well versed in the mash-up of politics, science, and media, ended up taking a similar position to Emanuel regarding advocacy, namely, that scientists are free to offer their research as expert advice but should stop short of policy "answers." Yet the line is much grayer than such a statement might at first appear. Providing scientific findings and "taking an advocacy position" that encourages action based on those findings are actions that scientists have often taken on contentious social issues. Indeed, important information may even require scientists to form ties with those using it both inside and outside government. Still, the tag "advocacy" seems to signal something completely different than this process, and reinforcing Fleck's concept of slogan words, it may be "degrading."

Cornelia Dean of the *New York Times* put it rather differently and perhaps closer to McCarthy's formulation than Holdren's. She considers scientists' contributions to public debate to be part of their obligation as citizens, but she also points out that scientists aren't necessarily rewarded for it.

> There still are too many scientists who think that their job is to make findings and report them in the scholarly literature, and that if they have done that, they have satisfied the obligations of citizenship. And I think they have an obligation of citizenship to participate in the discourse of the nation. And because their voices are, on the whole, missing, the quality of our public discourse is debased, especially nowadays when so many issues have a science component. But, you know, science

as an institution does not reward this kind of behavior. Not only does it not reward; it can punish it. So I think that has to change.

Dean teaches a seminar class at Harvard for scientists about the media, and after our interview she wrote a book about this subject as well. This thinking is part of her curriculum, and she further articulated this conundrum of how far and how much scientists think they can say.

What you will hear a lot of scientists say is that it's not our job to make policy, which is their way of saying, just as the journalist says: I'm not going to come down on one side or the other of whether or not human-induced climate change is for real. But scientists are going to say: I'm not going to come down on one side or the other, whether you should have CAFE [Corporate Average Fuel Economy] standards or whether you should do this or that. To which, I say: fine. But the people who are going to make those decisions in an ideal world would make them with the best available information that they could have. And the people who have that information are you, and you all should make sure that when these decisions are made, whatever the state of information is that you have is in the room at the same time. But even they—scientists—get very nervous about it, like they know they are approaching the arena of policy, the hair starts to rise on the back of their neck and they get agitated. I think that's because they foresee, probably rightly, that there will be criticism from their colleagues if they are perceived as being in the media too much, and too much is very little in the world of science.

This is a fascinating parallel to make between the professional norms of objectivity and independence in journalism and science. And yet what Dean is saying is that scientists should intervene in ways that journalists should not.

Dean and I attended a media panel at Harvard's Kennedy School where journalist Ellen Goodman had made the statement that no graduate students in science should be allowed to graduate without knowing how to talk to media.[11] When I asked her about it, Dean agreed with the statement and said that was part of the impetus behind her teaching a class at Harvard. I told Dean that I discussed some of these issues with graduate students at MIT who work on climate issues. In particular, I put it to one who was active politically and one who was not. The politically active student said he agreed with Goodman's statement, but he said he thought it would probably hurt him if his own political activity were a well-known fact in his department. And he said that if scientists are seen to be talking to the media a lot, it does hurt their reputation. The other student said he didn't think most scientists were going to talk to media anyway. He said that only a few elite scientists do that kind of thing. He saw any kind of media training or obligation to speak as kind of pointless "for the rest of us."

At the AAAS meeting in 2008, I heard a similar expression from a participant in one of the sessions who was talking about how he had "paid for it" among his colleagues when he had spoken to media. A funding officer from a major government agency spoke up and said: "I thought we fixed that. I thought we made it clear that this kind of thing is encouraged now, and scientists would not be penalized for speaking to media." The participant replied that this attitude about not speaking publicly or to the media was still prevalent, and heads nodded around the room.[12]

When I told Dean about the MIT students' response, she said that whether they want to admit it or not, they are "ambassadors for science," just like she is an ambassador for journalism. Not only that, but many scientists are funded by taxpayers through their education and professional lives, as McCarthy pointed out. Dean acknowledged that this is not necessarily the best relationship: being indebted and therefore required to participate. She returned instead to her views of citizenship and its obligations. And she said she understood the concern about being labeled an "advocate."

> I can see why they worry about appearing to be advocates or appearing to have adopted a position when in fact the only position a scientist can legitimately adopt is the position of skeptic because it [science] is an enterprise of skepticism, but that's not to say that you can't say, "This is what the data tell us." And that can start to look a lot like advocacy when the facts are heavily piled up on one side.[13]

Scientists being beholden to Mertonian-like norms, uncodified, fluid, and negotiable as they may be, is not necessarily new insight. What makes the climate situation that much more complicated is the fact that near-advocacy often means choosing a side or defending one's work as being on one side or the other, and that process begins as early as graduate school. One of the two graduate students I referred to above (the nonpolitically active one) was sucked into the vortex of public scrutiny when his first published journal article was picked up by a skeptical blogger and used as evidence against the veracity of climate change. The student was forced to put up a website stating his position on climate change and defending his work against those who might recast it in support of *their* position. This is increasingly what it means to do science related to climate change. It means openly subscribing to (or rejecting) "scientific consensus," usually by way of the IPCC, where as Dean puts it, "the facts are heavily piled up on one side."

Mediarology and Hostility

The late Stephen Schneider, a Stanford professor for interdisciplinary environmental studies and biological sciences, was well known for his expertise debunking nuclear winter and more recently for his work on climate change, including quite prominently with the IPCC. During the course of my research, I heard him speak to both scientists and journalists. He also participated in a climate change boot camp for journalists at the University of Oregon that I discuss earlier in the book. No stranger to writing his own op-eds, testifying before Congress, or speaking with reporters, Schneider developed a theory about the way media, policy, and science worked together. On his website, there is an extended essay on the topic, footnoting many of his own media and policy interventions, called "Mediarology." For Bud Ward's book *Communicating Climate Change*, Schneider provides a sidebar summarizing his theory.

Schneider begins by stating that in most "advocacy-dominated" stories, journalists usually report "both sides of an issue." But in science, "it's radically different" owing to the "spectrum of potential outcomes, often accompanied by a history of scientific assessment of the relative credibility of each possibility." Instead of adapting to such a major difference, journalists employ what they normally would in a legal or political setting and lock scientists into "one of two boxed storylines: 'We're worried' or 'It will all be okay.'" Schneider echoes these concerns. "Being stereotyped as the 'pro' advocate versus the 'con' advocate regarding climate change is not a quick ticket to a healthy scientific reputation as an objective interpreter of science. In actuality, it encourages personal attacks and distortions."

If the blogosphere is any kind of metric, this is an eloquent summation of exactly what has happened to many scientists, including most of those I interviewed, as well as journalists like Andrew Revkin who've become prominently associated with reporting on the story. When Schneider talks about journalists in this general fashion, it's clear he's not referring to Revkin, Dean, Richard Harris at NPR, or any others in the small cadre of prominent elite science journalists whose careers have been dedicated for decades to covering complex science issues. Instead, he's talking about the much wider coverage this issue receives through different forms and levels (local, regional, national) of media. That said, the Boykoffs' article "Balance as Bias" looked at leading newspapers like the *New York Times, Washington Post,* and the *Houston Chronicle,* all of which had science sections at that time, when it concluded that reporters were erroneously reporting as if the science was not clearly indicating the veracity of climate change. Richard Harris publicly challenged this conclusion at one workshop I attended citing the consistent and clear *New York Times* reporting done by Revkin. Yet despite the presence of some high quality reporting on climate change, the sense among most scientists I've talked to and that have been recorded in venues like the AAAS or Bud Ward's scientist/journalist workshops is that journalists have, by and large, erroneously forced scientists into pre-made categories along the lines Schneider describes in order to both simplify the story and report adequately on a much debated issue.

Direct address to publics without the help (and gatekeeping) of journalists has been aided greatly by the blogosphere and other online platforms. Some scientists like Michael Mann have fought back by establishing their own blogs. Mann has also written a book about his experiences "on the front line of climate change" (2012). Mann's research includes the much-maligned by skeptics and much-used by advocates (including Al Gore) "hockey stick" graph (featured in the 2001 IPCC report and later investigated by a National Academy of Sciences committee). MIT's Carl Wunsch was horribly mischaracterized as a skeptic by the documentary producers behind the UK Channel 4 film *The Great Global Warming Swindle;* he, too, fought back with a website statement. During my research, he did not respond to requests for interviews, but according to others, MIT was investigating legal action on his behalf. The graduate student I talked with whose research was similarly mischaracterized by a blogger, who was likely not a journalist, used the Internet as well to clarify his position. The Internet, then, has proven a "work around" to dealing with the conundrum Schneider presents.

Yet mainstream media remain a problem for many. Kerry Emanuel explained the problem as attributable to media, general scientific literacy, and scientists' desire to speak to a public somewhat informed about science.

> The problem with communicating to the public is that we want to assume that they have a rudiment in scientific reasoning, not that they're experts or that they're familiar with the jargon, but they generally know a little bit about science. And we're not happy or even willing to communicate with people who simply think that they shouldn't have to know anything about that. And unfortunately that does include a lot of reporters, whose backgrounds tend to be in the humanities where they've been taught to be hostile to science and scientific reasoning.

Echoing Boyce Rensberger's "simple machine," which I described in chapter 2, Emanuel told a funny story to illustrate his point.

> I once had a reporter years ago in my office and he was trying to get me to keep simplifying things way beyond what was reasonable and I finally—I think out of exhaustion and exasperation—said to him, "Well, it's like this: cold air sinks and warm air rises," and he paused and said, "Okay. Now, could you explain that in terms anybody could understand?" (laughter) The public and the scientists have to meet in the middle—maybe not exactly in the middle, but there is a level of ignorance which we simply can't deal with and that's true in any field.

Emanuel used the Weather Channel as an example of educating the public, noting that for more than twenty years, they've educated their viewers about satellite imagery and other meteorological elements because the viewers are interested and want to learn. It's worth noting that this is the work that social groups like ICC, Ceres, and Creation Care perform and even move beyond. They help to foster and meet an appetite to know, to learn, connecting climate change to existing forms of life, practices, knowledge, and beliefs.

When I put the notion of "hostility" to science to Cornelia Dean, she countered that science reporting has usually been too close to "cheerleading."[14] Harvard geologist Dan Schrag, who runs the Center on Environment, where Dean teaches, said he's had a "running argument" with her about who's at fault for the state of science reporting. Schrag blames media; Dean blames scientists' inability to communicate well. Schrag told me that Dean, who doesn't have scientific training, jokes she was made the science editor at the *New York Times* because she was seen one day walking around with a copy of *Scientific American* under her arm. (Dean confirmed this story.) Despite having a high regard for Dean's reporting, Schrag said:

> To me that's the problem. Would you hire somebody to be editor of the financial section who had never known anything about economics? It would never happen, right? Will you hire somebody to be head of the political reporting if they never actually had any credentials in reporting on politics? It wouldn't happen. But somehow with science, they allow people who have no education in science to suddenly then talk to scientists and report back. As if that's better because they are not biased by too much knowledge. That's pretty anti-intellectual. They don't say, "Oh, you went to the museum last week? Great, you can be the arts editor." It doesn't work that way. And in science news, it unfortunately is that way. If you asked how many science reporters have more than an undergraduate degree, and most of them aren't even in science, it's very small. But some of them are very good despite that. But you know that is the standard, right? That the newspapers do not, and the media, they don't value that.

Though such a critique might apply to Dean, whom Schrag clearly respects, it doesn't apply to former *Times* reporters Andrew Revkin or Boyce Rensberger, who both have science degrees at the undergraduate level. What Schrag's comment does raise, however, is the ways in which newsrooms have a much different norm for assessing who can articulate scientific findings for diverse publics, and the answer is not always someone who has a background or training in science.

Navigating Media and Policy Worlds

Schneider uses the courtroom metaphor to articulate what scientists are up against when they enter the public discourse arena. He makes only passing distinction between the arenas of policy and media—seeing them as they are, in fact, seamlessly intertwined—reinforcing one another, providing an echo chamber of responses back and forth. Yet there are distinctions to make between what each is dependent on and what incentives promote certain kinds of responses and practices. The role of money, pressures on existing models and structures, as well as relations to concepts of civic duty differ significantly. Media in particular and policy in similar but different ways work to craft and uncraft various meanings, promoting some and sidelining others. Schneider explains what he sees as the root of the problem in his online post, "Mediarology":

> The fundamental question related to climate change, then, is this: How can we encourage advocates to convey a balanced perspective when the judge and jury are Congress or public opinion, the "lawyers" are the media, and the polarized advocates get only twenty-second sound bites on the evening news or five minutes in a congressional hearing to summarize a topic that requires hours just to outline the range of possible outcomes, much less convey the relative credibility of each claim and rebuttal?"

One might suspect that Schneider would then call for a reform of the system of public address—the way Yale scientist James Gustave Speth did in a keynote lecture I attended at MIT in 2005. In the lecture, Speth castigated the media, yearning for the days when Walter Cronkite would have been able to elevate an issue like climate change. His solution was to find a way to talk directly to the public—through advertising, if necessary. Granted, media coverage has increased since Speth's lecture, but his suggestion of somehow going around the media speaks to the level of frustration.

Schneider, however, offers a path of navigating the system. He begins by asking a set of questions that gets at the underlying expectations of what function expertise is supposed to fulfill:

> Is there a solution to this advocacy-truth conundrum? On the one hand, it is an expert's responsibility to honestly report the range of plausible possibilities (what might happen?) and their associated (usually at least partially) subjective probability distributions and confidence levels. (What are the odds?) On the other hand, an expert may have a personal opinion on what society ought to do with a particular risk assessment. Can a scientist who expresses such value preferences about a controversial topic also provide an unbiased assessment of the factual components? This may be a feasible tightrope to walk, but even if one is scrupulously careful to separate factual from value-laden arguments, will advocates and advocacy institutions buy it as "objective"? An active effort to make our biases conscious and explicit via outside review is likely to help keep our science advocacy more objective. The more we discuss our initial assessments with colleagues of various backgrounds, the higher the likelihood of illuminating our unconscious biases, allowing us to better manage the "advocacy-truth" conundrum.

Here Schneider distinguishes between facts and values/opinions. When he spoke at the climate change boot camp in Oregon, he made a similar distinction, telling the journalists present to identify the differences when they pursued sources for their stories. For scientists, he recommends a path much like the one he's taken where values and biases are made explicit through what he terms a "hierarchy of backup products" that include op-eds, books, and popular articles that clearly distinguish between what's reliable and what's speculative. And he argues that scientists must not "abdicate the popularization of scientific issues" to those less knowledgeable or those wishing to simplify the science. He also recommends talking to colleagues of various backgrounds as perhaps a nod toward epistemic differences.

STS scholars have continually shown that the social is an inherent, constitutive part of the production of scientific knowledge, meaning that facts and values are always hybridized.[15] Jasanoff calls this the idiom of coproduction such that scientific knowledge is "embedded in social practices, identities, norms, conventions, discourses, instruments, and institutions—in short, in all the building blocks of what we term the *social*" (3). Bruno Latour's recent suggestion is that Science and the sciences are in fact two different beasts—the latter being completely aware of its contingent partial processes of gathering knowledge while yet, for the most part, often subscribing to the objectivity inherent in the concept of Science. Reading Schneider through Jasanoff and Latour, it would seem that he is indeed cognizant that the social always intrudes in practice, and yet there is utility in more traditional divisions and purifications such that expertise might have resonance with older ideals.

Part of that utility lies in the multilayered complexity of what it means to speak within the continually reorienting topography of American policy and media arenas—something akin to a vast network of complex stakes and alliances. Anthropologist and STS scholar Michael M. J. Fischer might describe this, using Donna Haraway's term, as "a cat's cradle situation" where every move affects every other element, reorienting the topography of the situation. Fischer (2003), writing about transformations in science and technology, describes this topography in his essay on emergent forms of life as one where institutions are mutating, new technologies are reconfiguring perception, there is massive economic and state restructuring, and new long-term risks prevail. Fischer argues that current modes of pedagogy and theory aren't able to address fully the questions of "heterogeneity, differences, inequalities, competing discursive logics, and public memories; complex ethics of advocacy and complicity; and multiple interdigitated temporalities" (2003, 39). The seemingly ungrounded ways of acting are emergent forms of life or a "sociality of action," replete with ethical dilemma, the face of the other, and historical genealogies, "requiring reassessment and excavation of their multiplicity" (58). Certainly, this is the case with climate change where reorientation, mutation, and veritable rabbit warrens of histories and meanings proceed, challenging those new and old to their layered applications at any given time. What emerges alongside, as Schneider's extended discussion online and many of the other examples in this chapter illustrate, are newly emergent norms and practices around what it means to speak for and about what climate change means—its form of life, and its associated facts, predictions, and risks.

Conclusion

This chapter explores some of the ways in which scientists have been pressed into near-advocacy positions on behalf of the veracity of climate science and the need to address the future consequences inherent in climate change predictions. Scientific consensus was intended to do some of this work, to move the questions from "Is it real?" and "Do scientists agree?" to "What should we do about it?" It's in this move that risk and uncertainty emerge as key terms of what Shackley and Wynne call a "boundary-ordering device" for the purposes of marshaling evidence and organizing research efforts such that an assessment of meaning, impact, and ethical implications might emerge.

Many have explained the negative responses to climate change as related to its uncertainty and lack of specificity, and this explains at least in part what has pushed the framing of "risk" further and faster. The risk framework allows for ethical questions to emerge from the range of possibilities posed by scientific findings, but "risk assessments" have primarily taken the form of cost-benefit analyses. Such efforts at quantifying risk elide the epistemological and ethical stakes in play, and they sideline other meanings and approaches to articulating and understanding risk and climate change as a form of life. Yet when instantiated within the vernacular of financial markets, risk can transform climate change into a rationale to act according to fiduciary obligations as chapter 5 shows.

While risk assessments and economic translations offer one avenue for analyzing how scientific expertise has been deployed in public discourse, science experts have often been called on to speak in more traditional ways—in response to weather events or on behalf of the work undertaken by the IPCC, for example. It is clear in many of the articulations and responses in this chapter that near-advocacy and the responsibility to speak about ethical implications sits uncomfortably with Mertonian norms. The spectrum of near-advocacy ranges from intense involvement in policy like John Holdren and Stephen Schneider or intense involvement with the IPCC like James McCarthy and Schneider to more spontaneous, event-driven involvement by Kerry Emanuel, Dan Schrag, or Naomi Oreskes. In trying to articulate why and how scientists should speak, rationales from both scientists and journalists utilize either historic scientific advocacy (by Einstein, for example) or citizenship and an obligation to taxpayers who fund research. But what ultimately compels many are the kinds of implications they see stemming from what their research reveals and the ethical responsibility to speak for such expertise.

What near-advocacy looks like when it reaches policy or media arenas where discourse and discursive strategies circulate and collide is not always predictable as the Senate hearings analyzed in this chapter suggest. And certainly as scientists, like journalists, enter the fray of public discourse via blogging and other media platforms, they will encounter other interpretive frameworks and epistemologies—other ways of making sense of the world. Hulme has suggested thinking of climate change not as a problem but as "an intellectual resource around which our collective and personal identities and projects can form and take shape" (2009, 326). And Hulme also advises that there are limits to how scientific expertise and findings should be deployed in public arenas. He suggests that (1) scientific knowledge is always incomplete, (2) science is shaped "to some degree" by the social processes in which it circulates, and (3) ethical judgments are "beyond the reach of science." For those who operate on various points of the spectrum of near-advocacy, points 2 and 3 may produce some tension. Near-advocacy is predicated on the notion that ethical questions emerge most forcefully from those for whom the findings are well known, and yet not all are willing to engage fully in the contentious public arena that awaits either at the Senate (depending on who is asking the questions) or the blogosphere on any given day.

While many in this chapter might take issue with Hulme's notion of climate change as "an idea"—rather than a fact in need of address, a problem in need of solution, or a risk in need of assessment—climate change does operate in public arenas like an idea. But it is one of many aspects of climate change's form of life. Climate change's meaning is understood through its use, and disputes over its use can also be understood to represent disputes over epistemologies and social order (see Shapin and Schaffer 1989, 15). It is this latter aspect which is most keenly at issue in the next chapter where climate risk acts to order financial markets and ways of thinking about and doing business.

CHAPTER FIVE

What Gets Measured Gets Managed

Every year in April, Ceres, a corporate social responsibility organization based in Boston, puts on a conference that by both academic and nonprofit standards is rather lavish. Attendees include representatives from Wall Street, pension funds, and socially responsible investing firms, as well as a long roster of small and large businesses that include some highly recognizable consumer brands like Johnson and Johnson and Ford. Well over five hundred attended the two-day event in 2007 and 2008—the years I was a participant-observer. What attracted me to conduct research with Ceres was their focus on capital markets and corporations and their work to transform climate change into "climate risk" for corporate America.

In 2007, the first session began with a breakfast discussion on the Global Reporting Initiative (GRI), a standardized accounting system for sustainability issues that scores companies on their performance on environmental issues, labor practices, human rights, and management strategies. Though Ceres is more established than Creation Care, the session had a similar feeling of being in the midst of an insurgency where the stakes were high and every comment reoriented what was up for discussion. Experiences and problems were raised, frustrations released, and opinions aired. The topic was not as much about GRI as it was about American capital markets in general. Representatives from major corporations like McDonald's and Office Depot traded experiences and insight with Ceres board members, socially responsible investors (referred to as SRIs), and nonprofit executives. They talked about the difficulty of integrating sustainability into business models, the moral rationale for doing so, and the stakeholder process that Ceres has established.

In the discussion, Wall Street's evaluation of their companies purely in the short term emerged as a serious and central concern. One contributor said: "Wall Street has the attention span of a gnat. It thinks long term is five years." Another participant responded in metaphor that they "hoped there would be a process of gently petting the dog [that is, Wall Street and capital markets] awake rather than kicking it." Others talked about the critical juncture confronting companies where what is at stake are communities and society. The key, one individual said, is recovering from "the hangover" of thinking that "if something is good for the environment, it must be bad for business," and vice versa. "The idea that it's a complicated process—I think we have to get away from . . . it's really about connecting people and ideas." The real key, another countered, was to look at things "in a more dimensional way." "The issues," claimed one major company representative, "are systemic. One company alone can't solve it, yet I haven't seen a model that reflects that." An SRI representative responded, "The reason we go after single corporations is that there's a domino effect," implying both a systemic strategy and effect to interventions.

Sitting in on this session felt a little like being dropped off in the deep end of corporate social responsibility debates, live, in-person, and with some of its key players. Participants in the discussion were wrestling on multiple levels with what it means for a company to be accountable, what conceptions of a just and productive society drive most corporate social responsibility (CSR) work, and whether these ideals could possibly be reconciled with one another. Ceres's much larger plenary sessions took place in massive hotel ballrooms and were considerably less philosophical, less full of debate, and less intimate. In the two years I attended, they included keynotes by the CEOs of Baxter International, Citi, Bank of America, Timberland, and State Street International, as well as noted activist and author Bill McKibben. Plenary panel sessions ran the gamut of putting Stonyfield Farms and Timberland together, two leading-edge companies on the topic of sustainability, while another memorable session dealt with how the public was embracing climate change; this included thinkMTV, Steve Curwood from PRI's *Living on Earth*, and the National Religious Partnership on the Environment. Breakout workshop sessions were held in medium-sized rooms that were almost always completely full to standing room only. They had titles like "The business value of sustainability," "The collision between coal and climate," "Global warming hits Wall Street," "The rise of ecomessaging," and "How insurance catastrophe models can help business and government plan for climate change."

During sessions, I found myself seated beside a Wall Street banker, several socially responsible investors, a highly ranked individual from a state treasurer's office, corporate representatives (from small, medium, and large companies), and environmental activists. Some attended because they were committed members of Ceres and others because they were being recruited to join. A few others were there because they were curious about how to tackle all of the environmental issues that were bubbling to the top, especially climate change. I had first heard of Ceres when they sponsored a talk at MIT by Solitaire Townsend, a UK public relations executive well known for her work on climate change communication. In Ceres's participation in the panel after Townsend's talk, it became evident that Ceres was approaching climate change much differently than the scientists in the room. The biggest difference was that Ceres often attached a dollar value to the investors they brought together for meetings as a way of heralding what those investors had at stake in relation to the risks associated with climate change.

Each year at the conference (and in most of the other events where I heard her speak), Ceres head Mindy Lubber took the stage to welcome guests and began by referring to the trillions that Ceres represents. At the 2008 conference, she referenced the $22 trillion in assets in the room, the $5 trillion represented in Ceres membership, and the $2 trillion represented in calls for addressing the climate crisis through Ceres's Investor Network on Climate Risk (INCR). It's from atop the trillions that Lubber commands the attention of the room and is able to talk convincingly to corporate audiences about "building sustainability into corporate ethic" and "managing footprint whether it be carbon, water, or labor practices."

She said that in 2001 Ceres began to talk about climate, and now "climate risk" is "as commonly used as sustainability." It's on this point that Lubber declared that Ceres and its members are "thought leaders" able to ask and increasingly answer what "sustainable governance" means. In the case of climate, if Lubber's October 2007 testimony to the Senate Subcommittee on Securities, Insurance, and Investment is a barometer, sustainable governance means standardized disclosure and regulation of the material and financial risks posed by climate change. This translation and transformation of climate change from being an ignorable environmental problem to being a pressing financial risk is a task Ceres has been working on tirelessly since 2001.

Ceres originally stood for Coalition for Environmentally Responsible Economies, but that's an explanation I was only able to find in old news stories about the organization. Some of those stories also note that Ceres is the name of the ancient Roman goddess of agriculture. The way Ceres staff generally describe their organization is as "a coalition of investors and environmentalists." Longtime Ceres employee and the director of special projects, Chris Fox, told me that the pairing is often a hard one for reporters to get their heads around. He said they usually want to pick one—Ceres is either an investor group or an environmental group. Indeed, it's not been a usual occurrence to find Wall Street types and prominent environmental activists at the same conference, talking about the same thing: the magnitude of global warming and the need for policy and solutions. In his 2007 plenary address, activist and author Bill McKibben paid Ceres one of the biggest compliments possible from a self-described "ardent environmentalist." He said, "Ceres is one of the few places I know that enjoys the trust of the entire community." He also pointed out that "we are at a moment where we can stand shoulder to shoulder on global warming with the corporate community." However, "shoulder to shoulder" does not mean that tension, productive or otherwise, does not continue to exist in such a coalition. Lubber in her conference opening speech in 2008 reminded the crowd that when it comes to addressing sustainable governance, "we must continue to dance at the intersection of our differences."

Despite McKibben's flourish, Ceres's work has not been greeted by the same fanfare as the Inuit human rights claim or the seemingly new turn for evangelicals toward climate change. Rather, over a twenty-year period beginning with the *Exxon Valdez* oil spill off the coast of Alaska in 1989, Ceres has labored to build a coalition that negotiates between major corporations, socially responsible and institutional investors, and environmental groups primarily through a process of stakeholder engagement. Ceres had more than seventy corporate members as of 2014.[1] They also labored intensively for over a decade to put GRI in place as an internationally used and recognized standard for sustainability accounting. In 2003, they launched the Investor Network on Climate Risk, an initiative that brings to bear the weight of numerous institutional investors to press for regulation and policy changes, as well as changes in corporate practices related to transparency and disclosure. At the 2010 and 2012 summits, INCR brought together 520 financial, corporate, and investor leaders.

Ceres has also championed and supported shareholder resolutions and produced many reports that score and rate entire industries and numerous corporations. While there are many groups like the Climate Group, USCAP, the Carbon Disclosure Project, and other environmental groups that do some of the same work, Ceres is the only organization that approaches its work as a coalition of environmentalists and investors and performs a range of tasks from shareholder resolutions and stakeholder management to producing reports and "convening" groups in order to move policy and change regulatory frameworks. Convening is the word Ceres uses most often to describe what they're doing with the annual conferences, INCR and its annual summit, and any initiatives they undertake to press for policy changes.

Science, climate or otherwise, is generally not part of this discussion. Rather, climate change and the disruption and possible chaos it will bring are considered facts worthy of translation into the vernacular of business. When I posed this to Ceres, the response from each of the people I interviewed was that there was no reason to delve into the science. "The jury is in," one employee told me. Before his appointment as a science advisor to the Obama administration, Harvard scientist John Holdren provided a brief at INCR summits, but more often than not, what Ceres has found is that investors are "not interested in the details" of the science and are not likely to be skeptics. Rather, they want to know what it means for their financial liabilities and risks as investors and corporate executives. Ethics are thus articulated in terms of fiduciary obligation, shareholder value, foresight, and insurance of an investment. In doing so, ethics not only dictate what actions are possible and desirable, but also situate climate change on a distinct epistemological terrain.

Climate risk is predicated not just on scientific facts but on the basis of structural considerations related to markets and corporate logics and grammars. In such a formulation, climate change's form of life is articulated with moral and ethical contours that are deeply etched with the notion of winners and losers. Climate risk thus becomes a metric by which competitive advantage might be established and evaluated. This isn't obviously a new turn for markets to adopt metrics based on social or scientific problems, but climate science produces probabilities that can be extrapolated into a variety of interdependent risk scenarios with financial implications. The assumptions underlying such formulations and weighting, coupled with capitalist imperatives, generate a distinct set of goals, standards, and ethical questions.

Navigating Vernaculars and Forms of Life in the Corporate World

Since at least the 1970s, scholars and activists interested in CSR have put some hard questions on the table for businesses to consider, questions that a growing field of scholarship has begun to explore: What's a business for? Can a corporation have a conscience? Is the social responsibility of businesses solely to increase its profits, as Milton Friedman argued (Friedman 1970; Harvard Business Review 2003; May, Cheney, and Roper 2007)? Climate change presents a set of questions that builds on these questions, but as "climate risk" illustrates, it also confronts notions of how risk is articulated, accounted for, and managed. This chapter specifically focuses on how climate and climate science have been translated into the vernacular of business through actions taken by Ceres. This chapter then asks questions about CSR discourse that makes both change and claims to it possible. What does climate change sound like when it gets translated into an economic rationale for acting through prefigured modes of corporate practices and performance measures? How is climate change factored into the discourse of investment, insurance, and risk? How is Ceres a part of the changing relationship between environmental activism and corporate responsibility? How should we account for aspects of corporate image, association, and branding that play a role in a decision to join Ceres and be part of a CSR-oriented dialogue?

Corporations have traditionally been seen as opposing environmental concerns and, in the case of major extractive companies like Exxon, have funded climate change skeptics and "the production of doubt" (Hoggan and Littlemore 2009; Lahsen 1998, 2005a, 2005b, 2008, 2010; Oreskes and Conway 2010). In seeding the impetus for action on sustainability issues like climate change, Ceres has built intensively on the success and assemblage of institutions and modes of interventions related to earlier anti-pollutant activism and anti-apartheid divestiture campaigns (Hoffman 1996). It has developed a number of tactics and strategies—notably, stakeholder management and shareholder activism—that navigate the proverbial terrains between stick and carrot. As evidenced by McKibben's affirmation of its strategy, Ceres has managed to extend its influence among corporate leaders as well as rank-and-file companies and maintain an integrity that warrants respect among environmentalists.

Ceres's utilization of the term *climate risk* speaks to the ways in which it has been able to seize upon the vernaculars inherent to the world of finance and corporate interests. Climate risk, perhaps more than the actionable vernacular advanced by Creation Care, reflects a wide spectrum of possible outcomes related to climate change. It advances notions of precaution as well as direct effects, drawing in part on quantifiable damages wrought by weather-related events (including Hurricane Katrina). This formulation can be critiqued for its overreliance on insurance rhetoric that ultimately benefits the insurance industry, but it also raises two other key issues.

The first is that risk is a modernist framework that works as both a herald of change and an impetus to more fully actualize the assemblage of institutions, modes of speech, and disciplining materiality (workshops, initiatives, briefings) that will address such change (Beck 1992). Risk frameworks imagine that change can be managed and accounted for, while also recognizing a spectrum of uncertainty. Crucially, too, the framework of risk and its attendant reordering of assemblages create new conflicts, inequalities, and political alternatives. These are the kinds of ethical problems that scientists, Inuit, journalists, and evangelicals have all been calling attention to in various ways (and in varied vernaculars). Yet in Ceres's formulation, there are not just losers. There is also an attempt to carve out exactly who is a winner and how businesses and investors might benefit from recognizing this risk now and not when it fully materializes. Not only that, but Ceres is part of a much larger effort by many groups to create market-based incentives that reward this forward-thinking recognition.

The second issue is that Ceres struggles with the short-termism or "liquidity," as ethnographer Karen Ho has termed it, endemic to Wall Street valuations (2009, 2010). Risk, as Ceres defines it, disrupts the usual straight line toward quarterly evaluation. It requires that investors and companies think in terms of protecting infrastructural investments and other vulnerabilities that would be encountered should the extreme end of climate change occur. But risk is also a double-edged term, as Ho illustrates, where a valorizing of risk has led to an unwavering belief that high risk leads to high rewards, undermining the very shareholder value that such risk-taking purports to increase. So even if climate risk were to be recognized as an underlying risk, when and how to intervene still depends on valuation and what's perceived as the best decision in profit-centric, loss-averse frameworks.

Ho further points out that before the 1980s, corporate and investment discourse spoke about much broader consumer, employee, and stakeholder engagement alongside shareholder value. But more recent discourse focuses solely on shareholder value—again, leading to the justification for risk-taking that leads to boom and bust scenarios. Ceres, too, articulates the need for action using notions of shareholder value and the fiduciary obligation to maintain and increase such value. Climate risk acts in Ceres frameworks to compel action on behalf of the shareholder, and its coalition of institutional investors and corporate leaders join together in Ceres-led and organized networks to demand such actions.

What bringing Ho's ethnography alongside my narration of Ceres's articulations illustrates is that Ceres must clearly negotiate and demarcate its use of a business vernacular that is at best muddied by multiple interpretations and translations within the multilayered corporate world of investment, capital, and management practices. Ceres uses terms like *risk* and *shareholder value* that are their own forms of life in order to mobilize actions and assemblages that will address climate change and utilize its risk factor as a means to effect greater change within processes of financial valuation and market practices. They often work by way of competitive advantage where some companies are doing it because their competitors or their industry as a whole are moving in that direction.[2]

From Valdez to Ceres: The Evolution of CSR Activism

Ceres was launched in conjunction with the announcement of the Valdez Principles, named for the *Exxon Valdez* tanker that ran aground in Prince William Sound, a remote area off the coast of Alaska in 1989. Spilling between 11 and 32 million gallons of oil (official estimates are widely considered to be too low), the Valdez spill became a rallying point for environmentalists and socially responsible investors who spent the summer following the incident formulating the Ten Principles: protection of the biosphere, sustainable use of natural resources, reduction and disposal of waste, wise use of energy, risk reduction, marketing of safe products and services, damage compensation, disclosure, environmental directors and managers, assessment and annual audit. They were renamed the Ceres Principles in 1992 because as a Ceres representative pointed out at the 2007 conference: "They used to be called Valdez until someone from the Audubon Society mentioned you wouldn't call Audubon the 'dead oily bird society.'"[3] Language, as I learned in my research with Ceres, is of particular importance to their organization and something they're likely to point to as a marker of success.

Success of any kind was hard to foresee when the Valdez Principles were first announced. The *New York Times* ran a story almost immediately in September 1989 titled "Who Will Sign the Valdez Principles?" It described the Principles as being a surprise to corporations, none of whom were involved in the drafting process, and quotes corporate representatives as saying that the Principles were either impossibly broad or already representative of what they were doing (Feder 1989). There even seemed to be some offense taken by corporations who were already trying to integrate progressive environmental action into their business. Joan Bavaria, first head of Ceres and a well-known SRI proponent with her own firm based in Boston, responded to the criticism by saying that the Principles were about trying to find a way to reward progressive companies.[4] She didn't negate the challenge inherent in such a task, noting that many companies might have an easier time filling out the thirty-seven-page questionnaire required to join Ceres than they would signing (and therefore promising to adhere to) the Principles. Environmental groups like the National Audubon Society and the Sierra Club stated that they would use the Principles as "a basis for exerting economic pressure, possibly including consumer boycotts, on companies that fail to address their concerns."

In a *Washington Post* profile of Bavaria the following year, she said she wasn't willing to go that far just yet. She had mainly been "using persuasion and the threat of shareholder action to gain signers" so far (Hinden 1990). The *Post* noted that Ceres was being lent a hand not just by environmental groups but by institutional investors like the California and New York pension funds and the Interfaith Center on Corporate Responsibility (ICCR), both of whom were already pressuring Exxon to sign the Principles.[5] In the same article, a manager at the U.S. Chamber of Commerce was quoted as saying that Ceres was "naïve" and that it wasn't that easy to put a "litmus test" to companies where they would have to say yes or no.

The Valdez/Ceres Principles were not the first to undertake such a test, however. They built upon groundwork laid by the seven Sullivan Principles that were introduced in 1977 to address corporate involvement and investment in apartheid South Africa. A 1991 scholarly analysis and comparison of the Valdez Principles with the Sullivan Principles found the Valdez comparatively "elusive and complex" without the finite geographical scope and time horizon (Sanyal and Neves 1991). It suggested there would be difficulties monitoring and enforcing the Principles because standards did not exist and performance goals were not legally mandated. The Sullivan Principles, Sanyal and Neves argued, were extremely effective due in part to the moral pressure exerted to sign the Principles, as well as the independent monitoring of compliance and the straightforward nature of the Principles that required corporations to comply or withdraw. The numbers back this up in rather stark terms. Between 1986 and 1990, 154 American firms ceased operations in South Africa, and more than $480 billion was divested. Sanyal and Neves concluded that it was unclear what "real rewards accrue to a firm that signs the [Valdez] code." The value of the Valdez Principles, they thought, might lay in its ability to assist in designing an integrated plan that responded to enduring public concerns about the environment, noting in particular that the public relations yield would likely be immense for any company that signed.

Ceres's first break into Fortune 500 corporations came in 1993. The Sun Company of Philadelphia, the twelfth largest oil company in the United States (in 1993), became the fifty-first endorser of the Principles and the first among its Fortune 500 peers. Sun negotiated some adaptations of the Principles. (A complete acceptance would have required them to go out of the oil business entirely.) In an interview following the ceremony, Sun's CEO said "he did not foresee major changes in company operations," since they had already been pursuing environmental initiatives (Wald 1993). Sun immediately embarked on a major advertising campaign announcing the partnership through full-page ads that cost more than 5 percent of their annual marketing budget, according to the *Philadelphia Business Journal* (Roberts 1993). The *Journal* surmised: "By being the first Fortune 500 company to endorse the Ceres principles, Sun has shrewdly positioned itself as a leader in corporate responsibility, an increasingly important image for companies selling to consumer markets." Yet for all the positive press, Sun was clearly aware of the scrutiny such a move would draw, as well as its binding commitments to Ceres to make data from its thirty-seven-page questionnaire publicly available, and to continue to monitor and report. Membership, too, came with fees that were indexed to a company's revenues—the high of which at the time was $15,000, according to reports. In 1994, Ceres's next big member to join was GM. The first major bank, Bank of America, joined in 1997.

The year 1993 proved to be pivotal not just in terms of membership for Ceres. *Pensions & Investments* reported that 1993 was the year "the number of shareholder resolutions on environmental matters" was higher than those on South Africa–related resolutions, "largely because of increased corporate environmental awareness and the prospect of the abolition of apartheid in South Africa" (Philip 1993). Apartheid formally came to an end in 1994 when free and democratic elections were held in South Africa. But a year earlier, the Ceres Principles had already begun to take the place of South African resolutions, with the total number of Ceres resolutions related to the environment going from thirty in 1992 to thirty-eight in 1993, and the total number of South African resolutions during the same period going from sixty-two to thirty. Besides Ceres requests, five other resolutions regarding the environment were put forward regarding ozone depletion, chemical emissions, and accident reporting.

Ceres resolutions requested that corporations either sign the Principles or that corporations engage in better reporting related to the Principles. These resolutions were put in front of major corporations like McDonald's, Bristol-Myers Squibb, R. R. Donnelley & Sons, Ford Motor Company, General Motors, USX-Marathon Group, Union Carbide Corp., and PepsiCo. (GM joined Ceres in 1994.) The groups that were reportedly most active on Ceres-related resolutions were the American Baptist Churches, the Evangelical Lutheran Church in America, the New York City Employees' Retirement System, and the General Board of Pensions of the United Methodist Church. *P&I* makes a point of noting that 1993 was also the year in which activists saw shareholder resolutions begin to make a difference: "Of the 253 [total] shareholder resolutions [including those related to the environment] introduced in 1993, 96 have been withdrawn to date because of agreements reached in negotiations between shareholders and management." According to *P&I*, 1993 was also the first year that activist groups began to argue that executive compensation should be tied to performance on social issues.

Through shareholder resolutions, Ceres Principles were used by activists in attempts to drive membership to Ceres and to make the environment a governance issue for corporations. Shareholder resolutions are heavily regulated by the Securities and Exchange Commission, but they are nonbinding. Primarily, they attract media attention and put pressure on companies by letting them know how much support there is for change on an issue without legally compelling them to do anything. They were extremely effective in anti-apartheid activism.

Ceres has benefited from and built on the legacy and infrastructure that anti-apartheid activism put into place. The very notion of calling corporations into account stems in part from this activism and particularly the Sullivan Principles. Many of the same entities who originated these methods like ICCR, pension funds, and other large institutional investors have continued to put pressure on corporations to account for their role in social and environmental problems confronting an increasingly global society.[6] Previously, many of the divestiture campaigns had been exactly that—getting rid of or avoiding investments in companies that violated moral or social values. This new era, however, ushered in a different mode of thinking, termed at the time by one magazine to be the use of "corporate dialogue" as a "tool for social change" (Klinger 1994). Corporate dialogue includes a wide range of "tools," from interviews and questionnaires to shareholder resolutions. Ceres can be seen as one outgrowth of this era of thinking. The participation of institutional investors like pension funds provided an added dimension beyond the usual suspects. They were interested in promoting social change as well as protecting their investments. The two, in their rationale, were inextricably linked.

Inside Ceres

The current offices for Ceres are located in downtown Boston in an old high-rise building, near the theater and shopping districts. It's a gritty part of town, especially in the summer when hot days hang heavy with humidity, and the subway vents built into Boston sidewalks only seem to intensify the heat and city odors. I visited the Ceres offices several times during the summers of 2007 and 2008, following their annual conferences in April, to interview staff members. Their offices are located on a floor with a couple of other NGOs. They've stayed in the same location despite doubling in size from twenty to forty staff members during the one year between my visits. Even with renovations to accommodate the growth, the offices are rather subdued and unadorned—almost like a start-up that may or may not stick around. It's a far cry from the sizable, stable presence the conference projects when it charges $400/person or more to attend. The conference, though, Ceres's Anne Kelly explained to me, is about (in this order) "convening," educating, recruiting, and reasserting the message of sustainability with people other than Ceres members. And, she said, it would be difficult or impossible to operate without it because it plays such a vital role in the ongoing work of the organization.

The history of Ceres (and of the CSR movement more generally) is clearly reflected in the histories of two of the key interviews I conducted. Chris Fox, director of special programs, described his years with Ceres in ways that match the overall history of the organization and changes taking place in the wider popular and business culture. He describes himself as becoming committed to environmental activism in 1989. He said that in the wake of the Valdez spill, there was a sense that "the government was not doing enough to improve company practices," and there was a "whole opportunity for a new strategy really to build on the success of the anti-pollutant movement." That strategy was to "harness the power of investors to focus corporate boards on how responsible behavior on environmental issues actually is good for business." He said it is modeled on this notion:

> There's three kinds of power people have in America: power as voters, power as consumers, and power as investors. And it's really the third power that I think is so untapped seventeen years later. People are still barely aware that they have this power. There's something like 50 million Americans that own mutual funds now, so there's a huge block of Americans that could be shareholder citizens (that's one term we use), that could be using their votes as investors to improve company practices.

Fox said he was inspired by the leadership of religious investors who "had tried to figure out how to fight apartheid using different shareholder activist tools." It's this interest in bringing together religious and environmental groups that eventually led him to seek a master's of divinity from Harvard. He mentioned Sister Patricia Daly, a nun, as a particular inspiration. Her work regarding Exxon is well known in SRI circles. Fox was the fifth employee of Ceres, and following my first interview with him, he was leading a media conference call regarding the latest round of shareholder resolutions to Exxon.

When I interviewed Anne Kelly, director of public policy, she introduced the other contextual element to understanding the work of Ceres—

the shift from a law enforcement–inspired approach to a dialogue-centric one. Kelly has a background in environmental law, and prior to Ceres she had worked extensively on forcing companies to integrate environmental regulation into their practices.

> I spent the first part of my career doing intense environmental enforcement—civil and criminal enforcement. I ran an environmental strike force with police officers, investigators, and engineers. I've done the photos and the undercover work. I used to have interns come and do sting operations. In the late 1980s, early 1990s, there was still a lot of open dumping, and you could plant people and get video. There was a lot of energy around using a lot of the techniques drug enforcement had used. Setting up fake companies and fake checkbooks and all that and putting that in the environmental area. Because, otherwise, the routine government enforcement had sometimes limited effect.

In 1995, Kelly took a break from this line of work and attended Harvard's Kennedy School of Government (KSG) to pursue a master's degree in public administration. After KSG, she worked with Mindy Lubber, who was then at the U.S. Environmental Protection Agency (EPA).

> At EPA, I started the beginnings of thinking—well, maybe you could actually sit down and work with companies instead of just trying to throw them in jail. So that was a real shift for me in thinking from the first part of the 1990s to the second part. There was a real shift from command and control to conversation, mediation.

She eventually set up her own firm, which did exactly that—mediation on environmental issues. She said she had always found the Ceres model compelling, but thought it required her to have an MBA. Once Lubber took the helm of Ceres in 2003, Kelly investigated and found she didn't need an MBA. When I first spoke with her, she was in the midst of wrapping up the work of her own firm and joining Ceres full time, though she had already been working at least forty hours a week for quite some time.

If the Valdez spill provided an initial event, the Sullivan Principles a precursor and model for corporate engagement, and corporate dialogue and investor "power" the underlying social trend, then it's the influence of socially responsible investing that provides the rationale and philosophy for thinking about changing the current system. Ceres does not bend toward anticapitalist inclinations so prevalent among many environmental and social justice groups—rather, it hopes to effect change by employing accountability mechanisms that value social and environmental elements alongside revenue/profit. Joan Bavaria, the cofounder and initial head of Ceres, personifies this approach in many ways. There have been three heads of Ceres. Bavaria's leadership was followed by that of Bob Massie and then Mindy Lubber.

Bavaria began much earlier to work on issues that are at the core of Ceres. In 1981, she cofounded the Social Investment Forum, also based in Boston, and a year later, Trillium Asset Management Company—one of the oldest and best known of the socially responsible investment firms. Ceres describes Trillium as "the first U.S. firm dedicated to developing social research on publicly traded companies" (Ceres 2008). When Massie, Bavaria's successor at Ceres, gave a talk at the Sloan Business School at MIT in 2007, he described Ceres as coming out of the Social Investment Forum and Boston-based investors who had used the Sullivan Principles in anti-apartheid activism—in both cases, a direct reference to the role of Bavaria.

In 2008, in the interest of focusing on and rewarding those who work on systemic change, Ceres created the Bavaria Awards for Building Sustainability into the Capital Markets—there are two: one for Impact and one for Innovation. At the initial launch of the awards during the 2008 conference, Bavaria gave a speech that provided some insight into her thinking. She said, "Wall Street and the market without steerage can wreak havoc" and "capitalism needs guidelines." With these statements, she made a direct reference to the ongoing subprime mortgage crisis of 2008 that had embroiled many. She spoke of the realms of finance, the social, and the environment as systems that needed to be thought of together, noting that "fiduciaries [for investments] must take into account the planet on which business feeds." She quoted Machiavelli's *The Prince*: "There is nothing more difficult to plan, more doubtful of success, more dangerous to manage than the creation of a new system. The innovator has the enmity of all who profit by the preservation of the old system and only lukewarm defenders by those who would gain by the new system." The awards reward those who work toward shifting the current capital markets from a "system focused on short-term profits toward one that balances financial prosperity with social and environmental health." In the initial year, 2008, the awards were given to those working on transparency and education.[7]

Ceres's Core Business:
Stakeholder Engagement and Sustainability Reporting

In its system and interventionist approach, Ceres seeks to demonstrate that "sustainability and profitability are not mutually exclusive," as one conference participant put it. Massie, the second head of Ceres, described the organization as an answer to the twin problems of a lack of political leadership on environmental issues and "capital markets as negative pressure" against actions taken on environmental issues. Kelly put it more succinctly in terms of strategy by describing Ceres as "using the leverage of the capital markets to influence companies." Massie said that Ceres's work is based on the adage "What gets measured gets managed; what gets disclosed gets done." In other words, Ceres provides a means and a suite of measures by which environment and social issues (labor and human rights, for example) can be factored into a corporation's overall strategy and valuation, both for its own purposes (employee retention or preparatory work for future regulation, for example) and for that of its investors (thereby avoiding and/or fulfilling shareholder resolutions). To do this, Ceres uses stakeholder management, standardized reporting and analysis tools, sustainability reports, shareholder resolutions, policy activism, industry reporting, and other levers to further its objectives. Organizationally, Kelly told me that they divide Ceres into work with investors and work with companies—at the time of my interview, she headed up the companies section, and Chris Fox led the investor section. These lines are hardly firm as they admit and as will become obvious. For the purposes of this chapter, I'm going to follow this schema by beginning with an overview of how stakeholder teams and reporting function, and then move on to how climate risk as a concept functions within the Ceres framework.

Ceres's members are either publicly held or privately held companies, environmental organizations, SRIs, institutional investors, and/or public service organizations. The latter groups (those that are not corporations) populate the stakeholder teams that help the corporation move toward sustainability goals. Essentially, stakeholder meetings put a corporation's critics at the table with corporate executives. When it works well, Kelly told me, the executives do less talking and more listening. I wasn't able to sit in on any stakeholder meetings, nor did any of the stakeholder team members want to talk with me about their experiences. They sign a contract that stipulates that what happens in stakeholder meetings remains confidential. From the outset, they decide how much to be involved and are paid for their involvement.

At the 2008 conference, I got some insight into how the stakeholder process functions and where its challenges may lie. The panel titled "Critical Friends: Engaging Stakeholders to Catalyze Change" first featured Tod Arbogast, Dell's director of sustainable business, and Ted Smith, cofounder of Silicon Valley Toxics Coalition (SVTC), who became a member of the stakeholder team once Dell joined Ceres. SVTC began when ground water contaminants were traced to high-tech development in Silicon Valley. SVTC engaged in protests and consumer boycotts against Dell—even at one point being accused rather overdramatically of "Hezbollah type tactics" because of a protest that involved Susan Dell, Michael Dell's wife. In other words, these are not individuals or organizations one would necessarily expect to be sitting on a panel together. SVTC was eventually able to explain to Dell, Smith said, that greening the supply chain was not just PR and that it would increase his market share. Smith said the way they achieved that was by putting pressure on Hewlett-Packard to disclose since they were perceived as the market leader in sustainability, and then demonstrated that consumer demand for greener products and recycling did exist. Eventually Dell hired Arbogast, and Arbogast said they chose to respond to the protests with "engagement." In media coverage, Dell does not give credit to SVTC, but Arbogast did not object to Smith's version of events while sitting on the panel (Gunther 2007).

The questions afterward were intense from the audience with some asking whether capitalism was up to "saving" itself, whose role it was to educate consumers, and still another who pointed out that sustainability does not always equal market share. Both parties readily admitted that stakeholder engagement hasn't necessarily increased the level of agreement, but it has changed the nature of engagement. Smith pointed out that Moore's Law states that technological innovation is exponential, but "the slope for sustainability is nowhere near as steep." The challenge then is to increase that slope, Smith declared. Ceres's solution is to do this through tools that promote listening to stakeholders and through reporting and disclosure.

The Dell discussion was followed immediately by Sandy Nessing from American Electric Power (AEP) and Andrea Moffat from Ceres. Moffat said that she and Nessing had met via a shareholder resolution and that she was initially surprised to hear from AEP. Nessing pointed out that AEP is still new to the issue of sustainability and that they were "caught between the duty to serve and protect environment and society." AEP, she said, was the biggest coal burner in 2007 and the largest carbon dioxide emitter in the Western Hemisphere, and they had initially opposed the Clean Air Act. But they had come a long way, as shown in their newly released sustainability report. She said AEP executives had never conversed with the array of stakeholders Ceres brought in. It became apparent during the first session that they did not know how to listen, and there was a lot of anger. Executives had "no idea about the perceptions and expectations of AEP."

When it was her turn to speak, Moffat described the executives as "shell-shocked" after their first meeting. She said they had a large table with eighteen stakeholder representatives and that Ceres had to turn other potential stakeholders away. She described the meetings as ones where the AEP executives sit at either end of the long table and the stakeholders sit in the middle and do most of the talking. Moffat said Ceres is not a "neutral facilitator." "Our goal," she said "is to get companies to increase exposure and transparency." Sustainability reporting is a tool that drives change in company strategy and structures conversations. She said Ceres was pressing AEP on a range of issues from policy positions on climate change and carbon sequestration and alternative energy to efficiency, environmental health and safety, an aging workforce, and coal in the supply chain. She said AEP has a new position on carbon, which she doesn't agree with, but it is on page 37 of their sustainability report—a report that board executives read and approved. CEO compensation is also tied to delivery on sustainability goals. Ceres's goals are focused on this level of integration, both at the board level and among the companies' executives.

Listening to these two cases of stakeholder engagement, one gets the sense that the process Ceres engages in is exactly that: a *process* of engagement with many twists and turns that reorient how and what goals can be achieved.[8] Moffat called it "a complex relationship with milestones" and suggested that all parties must have "realistic expectations." She said Ceres has to "sit back sometimes," and despite wanting to get to solutions, they have to "work toward prioritization." Julie Fox Gorte, a Ceres board member and SRI executive who moderated the session, described her frustration with other stakeholders who are not as focused as she is on dealing with carbon-related issues, so clearly there isn't a straightforward or unified process of directing stakeholder interests (Ceres 2010).[9] Nessing said they've had at least one stakeholder walk out and not want to talk anymore. Smith noted that there's an enormous amount of negotiation that goes on—on either side of the table. If anything, a minor criticism of this process might be that it is too incremental in moving toward tangible change. However, it's worth noting that whatever progress does get made occurs as a result of negotiating ongoing tension that does not negate direct protest action, but it doesn't directly enroll protest action either. So it takes a middle approach, and as in the case of Dell, it makes progress as a result of direct conversations that may or may not build on the pressure already applied through protest.

This is what results from what Kelly describes as Ceres's unique niche as "a convener" of diverse multisector parties in both the nonprofit and corporate worlds.

> There is an internal conflict within Ceres, a dissonance, that is really at the core of what we are. Because, on the one hand, we're supporting and partnering companies, and then, on the other hand, we're beating them over the head to take action. So we live with that and I think do a pretty good job of managing that dissonance and that, some would say, contradiction. Because companies don't really want shareholder resolutions per se, and we have a very active global warming shareholder campaign.

This is perhaps the best way to understand Ceres—as a suite of multiple pressure points where shareholder action, stakeholder teams, sustainability reporting, and other work Ceres does in the policy sphere and with investors work together to continually move forward, however incrementally, toward a different paradigm for management and accounting that includes material and financial risks related to the environment. What both Fox and Kelly emphasized in my interviews with them is that Ceres does not want sustainability issues to be relegated to a CSR department. Rather, they want it integrated from the board on down as a strategic and governance issue. This is part of what has made climate risk such an important tool for Ceres because it becomes, as Kelly put it, "a lens by which we see all the other issues." In this, there is a clear echo of epistemological statements made by evangelicals about how their beliefs inform their interactions with and learning about other kinds of knowledge, but here Kelly uses it as a call to action.

This view of climate change as risk calls companies to fully integrate a spectrum of concerns and possibilities that relate to massive environmental changes. Companies must account for their emissions as well as infrastructure and supply chains that might be vulnerable to such changes. These responsibilities, if they are to be fully addressed, can't be relegated to a part of the company. What Ceres generally recommends doing is establishing some kind of in-company committee that straddles different core areas of business, as well as developing some kind of accountability at the board level.

Greenwashing

A major point was raised by external critics of Ceres's inclusion of carbon emitters like AEP. Ceres, it was implied, makes it too easy to sign on to their organization and gain their "stamp of approval" and association similar to the initial concerns expressed by Sanyal and Neves when they compared Ceres to the Sullivan Principles. The inference of such criticism is that it would be better to continually hammer away at a company like AEP rather than reward them for any positive behavior with regard to the environment.

I was similarly surprised to see Suncor receive an honorable mention for their sustainability report. Suncor's involvement in Canada's oil sands would seem to preclude any reward for plans to move toward sustainability.[10] When I asked Kelly about Suncor in particular, she struggled to find the right words. She admitted that the extractive industries are perhaps the "hardest" and are the subject of internal debates at Ceres, but she said that they have opted to reward "best practices" for being best practices. And she concluded that being "at the table" even with the "highly unsustainable version" of a company allows Ceres to attempt to divert them to more sustainable practices. These internal debates point to the difficulty Ceres encounters as an organization in its role of wielding the stick and carrot—offering rewards or carrots through association and membership and sticks through shareholder resolution and stakeholder management.

The specter of what is often called greenwashing looms large when it comes to rewarding companies, even by virtue of associating with them. Greenwashing is defined quite elegantly by SourceWatch as "the unjustified appropriation of environmental virtue by a company, an industry, a government, a politician, or even a non-government organization to create a pro-environmental image, sell a product or a policy, or to try and rehabilitate their standing with the public and decision makers after being embroiled in controversy" (SourceWatch 2010).

In putting the greenwashing criticism to other Ceres and SRI representatives, several argued like Kelly that effecting small changes through membership and/or association is precisely the point. Joining is "sometimes the thin wedge that gets things moving," as one SRI representative said to me. In other words, if companies do start out by thinking it might be a good public relations move, they often get pulled in much deeper than they could anticipate and more often than not they begin to make real changes that may have a wider effect. Peyton Fleming, Ceres's senior director of strategic communications, echoed Kelly's perspective, but went further: "We don't feel there's much to be gained by working with companies that are doing everything right and are relatively small. We think the biggest gain could be from working with one of the largest companies in the business sector, and if you get them to change their practices, that will ripple through their industry."

Fleming said that's why they work with McDonald's, Ford, big power companies, Suncorp, and some of the largest banks. "A lot of people hate the companies I just mentioned, but on the other hand, we think there are things that you could point to within these companies that they've actually done a pretty good job at and hopefully they'll do better. So, yeah, we specialize in working with highly imperfect companies." That's not to say that investigation isn't required. An SRI team I spoke with said they rely on employees much more today to blow the whistle, which perhaps puts an enormous onus on personal ethics. In contrast, another SRI executive, whose mandate it is to invest in companies with reliable CSR practices, told me they have six people in their firm doing research on greenwashing and that they take the problem very seriously as a matter requiring investigation—not just happenstance revelations through whistleblowing.

Ceres goes one step further than reliance on either whistleblowing or research. They purport to build a kind of anti-greenwashing antidote into the structure of their involvement and ongoing relationship with corporations through the stakeholder process, stringent reporting, and a policy of pushing companies to remain competitive in their commitments to "going green." Fox explained it to me this way:

> In terms of our board, it's half environmentalists, half investors. So the environmentalists are always saying, "Nobody understands us and the urgency of climate change. We have to move quicker, we have to make productions faster, we have to get the world governments to agree." Well, there's a sense of impatience that we bring to it, and we

kind of challenge the investor and business communities to actually take bolder action than they have already taken, and for setting the standard for what constitutes responsible climate change behavior by investors and companies. We're constantly raising the bar, in other words, and saying the standard that existed three years ago is no longer enough. That puts us in a relationship of tension with big companies like Exxon, who think, "Okay, yes, they're doing the advertising campaign about how they care about climate change, so that's probably enough." And we're saying, "No, there's actually seven actions you should be taking to address climate risk that other oil companies are taking but you haven't taken." You know, it's our role to often help investors and journalists sift through what's greenwashing and what's actually responsible corporate action.

Shareholder resolutions, stakeholder management, and sustainability reporting work together, then, to accomplish the task of moving companies out of their comfort zone or public relations efforts and into responsible corporate action. Coziness, in a counterintuitive way, results not in corruption but in effecting positive social-oriented change. And as Fox points out, it's driven by the way Ceres structures itself such that forces within the organization remain in constant productive tension with one another. This in turn trickles down to how work gets tackled at the stakeholder tables, with shareholder resolution targeting and other initiatives that Ceres undertakes to promote CSR tools, accountability, and education.

It's Fox's statement that undergirds what I heard at the conference when the comptroller for the New York Pension Fund went so far as to call Ceres a "support service" for their initiatives. Ceres is a workhorse in terms of pumping out reports, supporting shareholder resolutions, and creating stakeholder teams, but its staff, as Fox pointed out, are also intent on strategizing how and when to push major companies forward. This is what keeps Ceres competitive as a CSR organization in a field of many approaches (Hoffman 2011). One of its key selling points is that it maintains enough independence to make an association with it valuable, and part of that valuable association lies in the fact that real progress is seen on issues related to that amorphous notion termed "sustainability."

Fleming added one more aspect to this value by pointing out how influential many of Ceres's reports have been in getting attention for issues (like climate) by relevant industries (like insurance and banking). He said they did this with climate change and banking responses to it, and then scored the actions taken (or not) by major banks, which in turn drove awareness to the issue. Banks then began to call and ask how they could build climate change into their business strategy.

> We don't really use the reports to sell ourselves directly, but it's mostly to identify issues ahead of the curve, and I do think there's some recognition that we are reasonably good at that. Abby Joseph Cohen from Goldman Sachs has said, "The reason I value Ceres is that they identify issues earlier than I would otherwise." So that's always a big challenge for us to try and stay ahead of everybody else in terms of tapping issues before they get a lot of attention.

How they stay "ahead of the curve," Fleming said, is by paying attention to "what investors should be caring about" and asking companies questions about that issue or set of issues. When I spoke to him in 2008, the issue Ceres was tackling was water scarcity. They later issued a report on climate change and water scarcity.

One of Ceres's key contributions as a "support service" has been the crafting of a standardized reporting system for companies doing sustainability reporting. Called GRI or the Global Reporting Initiative, it was launched in 1997 under Bob Massie's leadership, and it received a major boost in 1998 from the United Nations Environment Programme partnering with it. It was spun off from Ceres in 2002 and became a separate entity basing itself in Amsterdam instead of Boston.[11] At the 2007 conference, GRI was celebrating its tenth anniversary and the third incarnation of its guidelines, called "G3" for short. At his MIT talk at the Sloan School, Massie said they started GRI in order to craft something like the generally accepted accounting principles. He said the rise of the Internet at this time played an important role because of the low cost of international communication. There were (and continue to be) several competitive disclosure models, but none that covered all of the categories that Ceres does.[12] In addition, there were other scattered ways of addressing sustainability, and "every company had a pet NGO; every NGO had a bunch of pet companies." GRI provided unified reporting requirements for companies, comparable information for investors, and consistency and completeness for accounting purposes. By 2007, GRI had over a thousand companies that used its guidelines. However, Alison Snyder, the GRI representative at the 2007 conference, told me that only about one hundred companies in the United States were using the guidelines at that time, and none of them were Fortune 500. Most of the GRI use was happening in Europe and South America.[13]

Ceres's focus on climate change, in part, grows out of the successful launch of GRI in 1997 and its maturation over time until it was spun off in 2002. It was around this time that Ceres began to look for its next big project. Fox told me that they had started an energy and climate program in 2000 in order to "educate our network about the importance of climate change." But in 2001, Ceres representatives had a momentous meeting with Nell Minnow, founder of the Corporate Library and well-known investor activist. Minnow recently coauthored a textbook on CSR. Fox describes the meeting with Minnow this way:

> The meeting we had with her in September 2001 was influential because she cited these anthropologists who would go into Wall Street and study what Wall Street talks about and is obsessed with. *Risk* was the term they realized was the most powerful. So if you had to approach mainstream investors and said you should care about global warming in 2001, it basically would have led to the door being slammed. So we came up with the idea of linking a new concept of climate change to an old concept of risk and doing the scaffolding or whatever theorists call the ways that people learn new things. The other psychological studies on people in Wall Street is that the fear of losing money is actually more powerful than the greed of wanting to make money, and as fiduciaries and people responsible for other people's money, that's the biggest fear.

This was the genesis for transforming climate change into "climate risk," but Fox said that it wasn't just tapping into the concept of risk.

Ceres was also helped along by major events in 2001. First, the IPCC made its strongest statement to date with the third assessment report. Second, George W. Bush rejected the UNFCCC's Kyoto Protocol, and so the "scientific community's sense of urgency and the environmental community's sense of despair" at facing another four to eight years of no government pressure on corporations to account for climate change also drove Ceres to consider climate as a major issue to take up. Third, the Enron scandal broke that year—the largest U.S. bankruptcy until Worldcom the following year in 2002. It's this third element that created the biggest opportunity for Ceres to begin talking about climate change as a risk among investors.

> At that point [after the Enron scandal] investor anger at companies not being honest was at an all-time high, and there was a window

that opened that we put the concept of climate risk through, and said there's another risk that's not been adequately disclosed to you, and it actually has major financial implications for these companies and therefore your portfolios. You have a fiduciary duty to assess the financial risk posed by climate change. And we just asserted that and did our best to back it up with various lawyers and legal people. It was put into a study we did called "Value at Risk" that was done in 2002. So that was the first time really that the term *climate risk* was used widely, and we certainly popularized that term and kept repeating it, and now it's a commonplace term.

But again, it wasn't just raising climate as a risk with companies and investors and building on fears and/or new information. Fox said they purposely positioned climate risk as a corporate governance issue. "This became like the key that unlocked all these big pension funds because they had corporate governance departments. It just fit into a frame that they already got an approval for. Then it was not a social or environmental issue. It was a corporate governance issue." Environment, social, and governance issues, often shortened to ESG, are often put together and sidelined instead of being integrated into the core strategy of a company, but corporate governance for investors is a core concern.

Fox said they used a "two-pronged approach" that prioritized getting the attention of investors. First, they went after large institutional investors like California Public Employees' Retirement System (CalPERS), which Fox said was the largest public mutual fund in the United States. Second, they went "on a separate track targeting corporate boards," working with the theory that boards are "supposed to be accountable to investors." Corporations are usually structured in such a way that CEOs report to the board, and the board is accountable to the investors, hence the power of shareholder resolutions to shape major governance issues. Fox said the crucial breakthrough came when they got the U.S. treasurer and the California state treasurer to collaborate. They had already been working together on corporate governance issues related to the huge losses to pension funds stemming from the Enron scandal.

Fox said that together Ceres and the treasurers had an idea for a high-profile event. He said state treasurers are usually "political animals" and want to lead on the national and international stage, but there's a risk in acting alone. The treasurers suggested doing something at the United Nations in New York, and what came out of that suggestion was the Investor Summit on Climate Risk at the UN in November 2003. Sixty treasurers attended and called for the creation of INCR. Since then, summits have been held in 2005, 2008, and 2010. Attendees numbered 520 at the 2010 summit, representing $22 trillion in combined assets. INCR's membership (not everyone who attends the summit is a member) has grown from ten investors with $600 billion in assets to more than ninety investors with nearly $10 trillion in assets. Members now include asset managers, state and city treasurers and comptrollers, public and labor pension funds, foundations, and other institutional investors. What INCR points to as its many accomplishments includes promoting clean technology investments ($4.9 billion in low-carbon investments since 2005), publishing research for investors on the implications of climate change, improving corporate disclosure and governance, issuing a call for international leaders to pass a treaty on climate (signed by 181 investors in 2009), issuing a call for U.S. action in 2007 (called "Capital to the Capitol"), seeking mandatory disclosure regulation from the SEC, setting best practices for investors, and training investors on climate risks and opportunities. In 2010, it seems as if Ceres succeeded in at least one of these goals when the SEC announced that climate risk disclosure would now be mandatory (Johnson 2010).

On the website, INCR is listed as "a project of Ceres." When I talked to Fleming, who began work with Ceres after the formation of INCR, he said that funding also drove its particular incarnation. Fleming said that 75–80 percent of Ceres's funding comes through foundation grants. I was surprised to learn that fees from companies are not the main driver behind many of Ceres's projects and initiatives. Some of the foundation funding is earmarked for specific purposes—in this case, raising awareness about climate change as an investor issue.

Fox said their marketing strategy for the message of climate risk has been to just repeat the message: "That's what big companies do. They continue advertising their products." The implication is that climate risk as a discursive turn is actually *a product* for Ceres. The goal, Fox said, was to make climate change "a top tier issue for investors and business leaders by framing it as a work issue. That way they have to manage it." He said they try to do "a surround-sound strategy," first targeting investors, since companies can't ignore them. "That's the most important group of messengers to communicate with the corporate world." Ceres staff members identify board members and staff of large investors. One institutional investor I met at the conference who was not yet a member said they definitely receive regular calls from Ceres staff members. Ceres is limited by their own small staff of twenty-five (which grew to forty in 2008), but they do their best to do direct communication through phone, web, e-mail, webcasts, meetings, conferences, and media campaigns.

Fleming and Fox said that Ceres is primarily focused on business and financial press. Their coverage doubled and tripled year after year during the period I studied them, but their online presence has yet to become as robust as they would like. Fox's prediction was that as climate change increases in profile, "there will be much more specialization. Who's going to help me out as a consumer? Who's going to help me out as an investor? Who's going to help me out as a voter?" This kind of thinking about a proliferation of data such that different user perspectives will be required speaks to the changes that have occurred related to climate change and its attendant prospects.

"Just So You Know, We Love Exxon Every Day"

When I talked with Fleming about the popularization of the term *climate risk*, he said the *cost of carbon* is a similar term that has become common, but clearly it has much more to do with regulation as well. The cost of carbon refers to what the price is on carbon trading markets or a potential carbon tax on all fossil fuel usage.[14] The cost factor, Fleming said, is what is likely to change business practices, or at least that's the hope.

> Ultimately, we want to change business practices and Wall Street practices and how they start factoring these kinds of issues in, and we're now seeing banks and Wall Street firms evaluating investments and including a cost of carbon in their decision making. That's ultimately what we're trying to achieve. Just as a company or a Wall Street firm looks at interest rates as an issue that they should always be paying attention to, we also want them to pay attention to what the cost of carbon has gotten to be in this five-year investment or ten-year investment. And are you factoring that into the value of the company or whether or not it makes sense to build that project? Those are the kinds of actions we're trying to ultimately achieve because that will then move the markets to be more responsive to issues like climate change. So that's our grand vision: trying to move the capital markets to operate in a more sustainable way. You can only do that by creating institutional change. I mean, the ultimate endgame is to improve the

world and improve environmental conditions and such, but you can only do that if you get companies and investors to start building these kinds of metrics into their way of functioning.

At the annual conference, this kind of reasoning is couched in terms of what the investor wants, requires, needs. And it's kept at a rather high level much like this explanation. There is little mention of the problems with carbon development mechanisms (CDMs), whether or not emissions trading actually decreases emissions, or debates about whether "cap and trade" or a carbon tax are better options.[15] Instead, the focus is more generally on the transformative power of creating metrics that take into account climate change as a risk and opportunity for individual businesses and industries. In pointing this out, I mean to signal that Ceres makes tactical choices about how much to challenge companies and capital markets, as opposed to partnering with more basic critiques of the ways in which new risk paradigms have caused new inequalities and regionalizations to erupt.

This focus on instituting new metrics and augmenting the existing ones is particularly evident in reading through the many reports Ceres has released regarding climate change. New metrics provide a kind of ultimate lever for effecting change, and it goes back to something Massie said early on in my research: "What gets measured gets managed." Mindy Lubber uttered the same phrase in one of her conference addresses. Metrics offer a way of quantifying (and rewarding) success, and expanding current metrics is meant to trigger a new set of practices in order to meet their demands. Institutional change results because of the new practices and thresholds set for their related metrics—for example, the integration of clean energy technology or the reduction of greenhouse gas emissions.

And yet one of the clearest complaints I heard at both conferences is that Wall Street doesn't get it—"it" being sustainability and/or climate risk. When I asked Anne Kelly what that meant, she said those complaints were referring to "the absolute embracing of short-termism, the absolute insistence on measuring everything by a quarter." At my first breakfast session at the 2007 conference, this was the subject of the meeting, and it wasn't just about time frame, but the nature of the measurement as well. The discussion participants talked about the fact that civic society spoke a "different language" than Wall Street, and companies were "beholden to analysts," who don't understand what the company is trying to do. One of the participants bluntly asked: "How do we create opportunities for companies to do something differently?" At another session later in the day that focused on how Wall Street finally might be coming around, one of the statistics an SRI contributor mentioned was that hedge funds tend not to hold stock for more than sixty days, so that shortens the window even more than quarterly earnings reports. This is the central issue confronting companies that seek to make changes or investments in sustainability-oriented goals, which usually require much more than a quarter to see results or returns.

Kelly explained what they're up against with Wall Street analysts: "Chris Fox did an interview with CNBC six weeks ago about Exxon Mobil. Chris was saying that Exxon's lack of address of shareholder resolutions and climate change issues is completely unacceptable. And the guy said, 'Just so you know, we love Exxon every day.' And that was the quote from the Wall Street guy. And if you look at the way Wall Street rewards, why do bad companies continue to do well? Because Wall Street rewards them."

Exxon continues to be "loved" by Wall Street because its earnings consistently rank in the number one place or right behind it (often battling with Wal-Mart) on the Fortune 500 company list. It has been a continuous target of Ceres and CSR shareholder activism, but Exxon's directors had refused to yield at that time despite direct pressure from the Rockefellers, whose family founded Standard Oil, of whom Exxon is the current incarnation (Carroll 2008). Kelly mentioned that looking at Exxon is a good test for showing how much work Ceres still has to do, and this is an instance where their efforts have failed to make substantial changes. Yet shareholder resolutions are generally as much about directing change within an organization as they are about garnering attention from media about the issue the resolution is putting forward. Such a strategy with Exxon is not entirely without merit or results then. Similar to the other social groups researched here, notably the Inuit human rights petition, it attempts to *change the discourse* and, in so doing, expands the notions associated with climate change as a form of life, establishing new starting and referent points and enrolling new players, institutions, and forms of expertise.

At the 2007 conference session on how Wall Street might be coming around to climate change as a pressing issue, one of the presenters put up a slide that showed all of the major investment banks—the majority of whom are encapsulated in the term *Wall Street*—have begun to do something about climate change. One of the prime instigators has been the carbon futures trading that has gone on in Europe in relation to the

Kyoto-based emissions trading. Some like Goldman Sachs have begun to issue reports and speak out on the subject. One of the participants at the conference session said, "The race is on on Wall Street" to move on this issue.

At the conference, I met the then new environmental director at Morgan Stanley, Jim Butcher, and later interviewed him in his Manhattan office. Morgan Stanley had a large unit that does carbon trading based in London, but was also looking to keep up with the pace being set on Wall Street. Butcher came from a scenario planning and consulting background, and he said Morgan Stanley was the only bank on Wall Street to have an internal scenario planning unit. He said the business environment was changing, and they needed to "get on board." He said they were spurred on by actions taken at Citibank and Goldman Sachs, but the *Stern Review* (examined in chapter 4) was also a key instigator of concern. He said that in 2006, climate change was not "on the forefront" of executives at Morgan Stanley, but that had changed. Butcher's role was to review what Morgan Stanley was already doing and engage in a broad stakeholder process. Risk and opportunity were something he saw as going together in part because of his scenario planning background. He said at the time of our interview in 2007 that climate change was not yet woven into research, nor was it consistently tracked, and he saw that as a weakness. Butcher said he saw himself as a translator of different perspectives and that bringing the science together with the intensive "language of financial services" was part of his role. Butcher's commitment to acknowledging climate as a risk reflects much of what Ceres's mandate and work have been about, but criticism emerged the following year that drew Ceres's methods into question.

In 2008 two Dartmouth College professors, Karen Fisher-Vanden and Karin S. Thorburn, produced a report that quickly became news. They studied the stock performance of companies who joined Ceres and Climate Leaders, an EPA industry-government program that mandates greenhouse gas emissions reduction. They found that companies that joined Climate Leaders received a negative reaction. The *New York Times* quoted Thorburn as saying "The pattern was clear—the more aggressive the goal, the more the stock price fell" (Deutsch 2008). Ceres membership, on the other hand, came with no significant reaction. Fisher-Vanden and Thorburn concluded: "The stock market is saying, don't count on voluntary initiatives." In response, Mindy Lubber, Ceres's board member Julie Fox-Gorte from Pax mutual funds, and representatives from Goldman Sachs and Morgan Stanley all agreed that the best run companies are those that also perform well by environmental metrics. Lubber, in particular, wouldn't accept the finding that companies get dropped because of environmental initiatives. Otherwise, she argued, "you wouldn't see so many companies addressing climate change with such a vengeance" (Deustch 2008). Quoted in the same article, Fox-Gorte also rejected the study's findings, saying they were "measuring the most ultramyopic reactions" that reflect short-term thinking.

The 2006 Ceres toolkit for corporate leaders that deals with "Managing the Risks and Opportunities of Climate Change" is also an attempt, like Butcher's role, to bridge the language of financial services and provide a roadmap for change. It makes clear how the Ceres methods fit together when it says, "Most successful corporations engage with concerned stakeholders, disclose their strategies to investors, and take concrete actions to manage risk and capitalize on opportunities" (2006, 2). It suggests ten steps to developing a comprehensive climate change strategy, grouped into a diagram with three overlapping circles entitled "assess," "engage," and "implement"—with the word *disclose* set around each overlap of the circles. The first four steps include creating a climate management team and a board oversight committee, measuring greenhouse gas emissions, computing physical, regulatory, and financial risk exposure, and assessing strategic, branding, and product opportunities in relation to climate change. Steps 5–7 involve creating plans to reduce emissions and risk. Step 8 suggests engaging in policy dialogue about reducing climate risk and enhancing opportunities. The final two steps are public disclosure and engagement. AEP, GE, Ford, Chevron, and Bank of America are used as exemplars of best practices on various steps. This is another key tactic of Ceres—getting industry leaders and major corporations on board.

What becomes evident reading this toolkit (as well as reports prepared on the banking and insurance industries) is that it is not just about getting corporate leaders or large corporations to take climate change seriously enough to change their practices and R&D investments. It is also about the power of association. Certainly, Ceres benefits from an association with these major industry leaders who are looking to make a change in their public image or to intervene in nascent policy debates early on. But it's not just an association with Ceres. The exemplars listed in the toolkit have begun to distance themselves from certain attitudes about climate change. Fox said that with apartheid, at some point it became morally repugnant to be associated with it. Similarly, he felt that the year

2005 marked a turning point for climate change because of several factors far outside of Ceres's control that were of immense benefit to INCR and the summit.

> I think 2005 is a critical year. Because of a variety of things that happened. Russia ratifying Kyoto in February. That was the moment where carbon was going to have a price. Industrialized nations were definitely moving ahead with regulating and then smart businesses like GE realized it was going to be a new market for low carbon products basically. Our second investor summit, which was much bigger than the first—twice as big—it was on May 10. And GE announced its Ecoimagination initiative on May 9, and then the whole buzz was about how the biggest company in the world had just announced this.[16] So that was the moment where the business community shifted and the kind of Exxon Mobil-dominated business world of businesses funding different industry associations and climate deniers and skeptics and all that—it just became irresponsible.[17] Because, if you really cared about your shareholders, you'd be figuring out how to protect and enhance shareholder value in a carbon-constrained world because that's now what we live in. It was like the moment where we just all acknowledged we live in a carbon-constrained world and that was going to be the future. And it was inevitable, and it was going to affect your market.

Kelly, too, agreed that this period was critical to the generalized sense of a sea change on attitudes toward climate change. Yet she questioned whether it is a matter of trickling up or trickling down. Shortly after the Kyoto ratification and Ecoimagination, Fox and Kelly both noted that Hurricane Katrina hit, which for many Ceres members, both new and old, was an event that reinforced the idea of climate change as an unavoidable material and financial risk factor.

Hurricanes Katrina and Rita—the images of destruction and suffering in particular—have been deployed by a range of actors to make climate change and inaction in the face of what could be much more widespread global suffering morally repugnant. The images, for those with or without institutional memories, are intended to create an effect not dissimilar to the years immediately preceding the fall of the apartheid government in South Africa when the success of divestiture and accompanying moral judgment was evident. What these kinds of moral prescriptions do in the face of risk is perhaps less easy to determine in stark terms of success or failure. The fraught process of creating and fostering a bandwagon for addressing risk, however, inevitably leads to a reorganization of some kind within affected institutions, their practices, and social groups.

Climate Risk and Recession

The framework of risk speaks to a certain view of economics, the role of corporations, and notions of social responsibility and to the larger context in which media, politics, and bureaucracy are a part of the formation of what Beck (1992, 2002) calls a "world risk society." Risk and the specter of catastrophic danger may act to unite societies, but in their demands that nations, or in this case corporations, unite and negotiate a response to the looming crises, they also create new conflict and political alliances, reorienting topographies of power, wealth, and capital, leaving in their wake new regionalization, inequalities, and exploitation. Positioning climate change as a risk implies that it can be rendered "predictable and controllable" in a modern society, and institutions, bureaucracy, and media rise to meet the challenge entailed therein. Risk thus works as a motivator precisely because it both heralds oncoming, unstoppable change and calls it into being. This is the paradox of identifying climate change as a material and financial risk and as an attendant suite of potential chaos and actual opportunity simultaneously.

This is not to say that the risk is not real. Perhaps one of the most staggering metrics Ceres has popularized through its report on the insurance industry is that the damage caused in 2005 by weather-related events rose sharply to $80 billion worldwide, equivalent to four 9/11s. Others at the conference and at Ceres have thrown around more colloquially the notion that Katrina incurred costs three times the cost of 9/11. At the conference in 2007, one of my favorite moments occurred during the panel that was discussing whether or not Wall Street was taking climate change seriously as a risk factor. One of the environmentalists got up and said, "Katrina blew the door down, and Al Gore walked through it." Bill McKibben later said something similar in his plenary address. "Al Gore" is a reference to the role of his phenomenally successful and Oscar-winning film, *An Inconvenient Truth,* released a few months after Katrina hit in 2006.[18]

I have used this phrase about Katrina and Gore many times in interviews with the people I've talked with for this book, and usually it elicits a smile. But generally a verification of the Katrina + Gore equation as the experience one had during the pivotal years of 2005 and 2006 depend on the importance ascribed to popular culture and trends in public opinion polling. Corporations (and, by extension, those involved in CSR activism) are incredibly concerned about consumer trends, competitive advantage (what other corporations are doing or in what direction their industry is heading), and maintaining investor confidence and shareholder value—all of which to greater or lesser degrees have some relation to public opinion. When Ceres talks about climate risk, these are exactly the levers they lean on, and events as well as popular culture in 2005 and 2006 finally began to provide some support for their claims.

I tentatively called 2007 the "summer of love" after I went to the Ceres conference because of the effusive tone of the participants, many of whom had waited for a decade or more for climate change to be taken seriously. This was a period whose highlights include a Supreme Court ruling saying the EPA could regulate emissions under the Clean Air Act, polar bears being listed as endangered species due to climate change concerns, insurance companies and Wall Street investment firms issuing high-profile reports about climate change, retired army generals issuing a report making climate change a security issue, and the release of the fourth IPCC assessment report with more dire warnings than the third. But the 2008 conference lacked much of the ebullient tone of the previous year. Instead, it was marred by the now epic subprime crisis, the roots of which continue to hamper economic recovery in the United States and many of the world's economies.[19] Citibank was deeply embroiled in the crisis, and yet its CEO, Michael Klein, still agreed to give the keynote at the Ceres conference in 2008. Klein began his keynote by saying that they continued with their commitment to focus on climate issues throughout the crisis because it's "so deeply integrated into our company that we couldn't pull back if we wanted to." Citibank set targets of reducing greenhouse gas emissions by 10 percent by 2011 from 2005 levels.

Still, I wondered whether or not climate risk could stand the test of a major economic recession, and I asked Kelly what they were saying about climate in 2008. She said they were positioning climate change as "another risk" that investors and companies might not be fully aware of or prepared to face.

> Well, we had used the subprime as a proxy for climate and Mindy Lubber has quite a good phrase around this—climate risk could be the next subprime crisis. We didn't pay attention to subprime for a long time and people kind of knew that that was building—they knew that sloppy decisions were being made. They knew that risks were being

taken, but nobody got on top of that, and then look what happened. And so she sometimes characterizes climate change in the same way, as a hidden risk that needs attention.

Looking back at how Ceres used the Enron scandal in a similar fashion, it certainly has echoes of the same logic. Crisis then doesn't have to work to supplant the groundwork laid by Ceres. Rather, it can be the impetus for further development of risk preparation and awareness.

Bob Massie called climate change a "disruptive syllogism." The *Oxford Dictionary* defines a syllogism as "an instance of a form of reasoning in which a conclusion is drawn (whether validly or not) from two given or assumed propositions (premises), each of which shares a term with the conclusion and shares a common or middle term not present in the conclusion." Massie explained that climate change is the largest physical change affecting certain industries and regions. It is hence embedded in every investment portfolio. Therefore, fiduciaries must assess what this change means for the investments they manage. Failure to assess equates to a breach of their fiduciary duty. This is the logic that pervades much of Ceres and SRI efforts to raise the issue of climate risk. But syllogisms often have important exceptions that can cause their seemingly perfect logic to unravel.

This is where Hurricane Katrina and the research that Kerry Emanuel released in the months before it hit become important verifiers of such logic-based assumptions. Kelly and Fox both pointed out that Ceres benefited enormously from studies that were already under way first when Katrina hit and then when the banking crisis hit. Fox said that Ceres was putting the finishing touches on their report on the insurance industry when Katrina hit, and Kelly said they had recently released their report on the banking industry when the subprime crisis began to unravel. Although the subprime crisis had nothing to do with climate change, Kelly said there were significant correlations that allowed them to capitalize on the risk paradigm: "We looked at their relationship to climate and energy and climate governance, and HSBC was number one with its 100 point score. I think they had 90 or something, and the lowest point-getter got zero points, and that was Bear Sterns, and we all chuckled. And there was a fair amount of publicity after that. It was just a coincidence that we happened to be looking at banks this year." The implication was that Bear Sterns was not a forward-looking bank concerned about the presence of risk in any of their portfolios, whether it had to do with mortgages, credit, or climate change. Bear Sterns collapsed in 2008 related to its role in the subprime mortgage crisis.

If we follow Beck's formulation through then, Bear Stearns becomes an example of the reorganization and reordering of new assemblages and institutions. Wealth in this case was not enough insulation from risk. Alternately, Karen Ho's explanation here might be that such risky behavior is endemic to Wall Street, creating more and more appetite for risk such that it unraveled the entire company—and could also potentially unravel much larger swaths of the capital markets and the American economy for years to come. Risk, in other words, begets yet more risk. And certainly, as Beck also predicted, activism like Ceres's work becomes important in identifying such risks—both the short-term evaluations of Wall Street and the possible long-term chaos associated with climate change predictions.

Insuring (and Educating) for Risk

The insurance industry has been on the forefront of concern about climate change.[20] At the 2007 conference, representatives from Swiss Re, F&C Management, and Fireman's Fund participated in a panel on how they are managing climate risk. The *Stern Review*'s findings were at the forefront of their concerns because of its estimates about how much mitigation and adaptation would cost if actions to stem emissions are not taken now. Alexis Krajceski from F&C summed it up when she said that for insurance companies, it's about accurate risk assessment as well as keeping insurance affordable and that they were counting on the wealth of "intellectual capital" they have with complex modeling to see them through. Swiss Re's representative, Mark Way, began his presentation by saying that Swiss Re "first labeled climate change an emerging issue in 1996." He said that average losses are increasing and the individual burden doubles every decade due to the demographics and economics of coastal development. This was the primary concern for the entire industry, Christopher Tulou from the Heinz Center concluded. Heinz and Ceres have been working together with the insurance industry on climate risk issues with their Resilient Coasts Initiative. Tulou said, "If there is a frontline, it's our coastal communities" who generate most of the country's GDP and contain two-thirds of its population. It's impossible to abandon these places, and so the "focus should be on protecting investments that are already there."

In my conversations with Kerry Emanuel, he similarly expressed concern about coastal development. He said that the federal and state governments were already subsidizing or providing insurance for communities in these vulnerable areas because insurance companies were unwilling to or their rates were prohibitively high. Conversely, Harvard's Dan Schrag, whose research focus is not on hurricanes, expressed a certain amount of skepticism about the insurance industry being involved in climate issues. He said that the insurance industry benefits enormously from something being labeled a "risk," because then they've got one more issue to insure. When I put this to Kelly, she said that she would think more about this, since it was the first time she had heard such a criticism. As she talked, however, she noted that the losses in the industry were very real, and she thought these business concerns were still the primary motivating concern for action on climate change.

The concerns of the insurance industry, however, echo, perhaps in starker and more concrete terms, the long list of risks Mindy Lubber laid out in a speech she gave at the Ethical Corporation conference in Boston in 2007. She said that companies are facing (in this order) regulatory risks, physical risks, reputation risks, and litigation risk. Krajceski clearly agreed when she said that she thought it was a "real opportunity to be involved in public policy debates" and that "insurance companies have much to gain from deliberate action and everything to lose from not participating." In the question period following various presentations, a representative from Marsh Insurance in the audience was called on to answer a question about policy changes. He noted that the EPA ruling, the latest IPCC reports, and the possibility of litigation were all on the horizon. He said they were "trying to elevate the level of public conversation" in order to account for these "tectonic shifts" in thinking about regulation, physical risks, and the need for disclosure.

In an effort to do just that, Marsh joined in a partnership with Ceres and Yale University to begin educating boards about how these issues were affecting corporations.[21] Kelly headed up this program on behalf of Ceres, and she considers this an analog to INCR, but one that deals expressly with companies rather than investors. Called the Sustainable Governance Forum on Climate Risk, it is described as a "leadership development program designed to help . . . address the problem of climate risk." Kelly told me that it brought together scientific, legal, business, and insurance aspects of climate change in an attempt to help corporate directors integrate it into company strategies: "Ceres figured out wisely in

2003 or 2004 that this is a top-down issue. That it's really important to train the boards of directors and CEOs. These issues are so big. They're a matter of long-term value, and they shouldn't be tucked into the ghetto of the EH&S [Environmental Health and Services] department. It's a matter of risk. It's a matter of long-term risk." This is a message that was often repeated at the conference and in other speeches I've heard Mindy Lubber give—that climate risk must be accounted for at the strategic level, and sustainability concerns must be integrated throughout the company. For companies, then, it must come from both the board and the executives in order for the kinds of change to occur that Ceres expects.

Kelly said directors don't like words like *teach* and *educate* because "they feel like they already know everything," so Ceres uses *engage* and *convene* instead. The forum describes itself as highly exclusive, invitation-only, "intimate," and "collegial" with a "discussion-based format." Part of the reason for both the format and the need for "engagement," Kelly said, is that corporate directors in the United States are an extremely "homogeneous" group—usually older, extremely wealthy white men. And while the climate risk framework has gotten through to some, there is a wide spectrum of knowledge. Kelly told me that she had one director come up to her after a session to say: "What's that word they were using? Starts with anthro." She said, "Anthropogenic?" And he said, "Yeah, I don't know what that means." She said that even within the Ceres membership, this represents one end of the knowledge spectrum while Seventh Generation, Aveda, Interface, and other highly progressive companies represent the other end.

The launch of the Forum was made during a session of the Clinton Global Initiative, and former secretary of state Madeleine Albright is on the advisory panel. The booklet describing and announcing the program has several bold quotes from corporate leaders, scientists, and political leaders. One from the *Wall Street Journal* stood out to me. It said:

> The group U.S. Climate Action Partnership (USCAP) stressed that by proactively dealing with the issue, companies can earn a voice in planning policy and thus avoid "stroke of the pen" risks in which new government rules can undermine a company's value overnight.[22] "If you're not at the table when these negotiations are going on," said James Rogers, Duke Energy's chief, "you're going to be on the menu." (Ball 2007a)

These are certainly the kinds of statements that motivate corporate executives and their company's directors, but they also are exemplary of the kinds of work that climate risk is doing for Ceres and others who have an interest in making climate change a CSR issue. Climate risk encapsulates a wide range of risks from regulation and litigation to material and competitive—all of which provide significant strategic concerns for both investors and companies.

Conclusion

What this chapter records is the work Ceres undertakes both with companies and investors in order to bring about a discursive shift and institutional transformation in corporate America. Climate risk is the latest and perhaps strongest of heralds that Ceres has been using to bring about such a transformation. Businesses and business media have, in recent years, often talked in terms of "going green" or becoming more sustainable. But it's not always been clear what this means in concrete terms—or indeed, whether it means anything at all. Ceres attempts to move beyond both the morass of what sustainability might mean and the problems inherent in the threat of greenwashing by asking corporations to become accountable to a set of stakeholders that are usually composed of nonprofit members of Ceres. It's through this mechanism that companies begin a process of identifying how they will become more sustainable across a number of markers, including how they will respond to climate risk. GRI, the sustainability accounting system that Ceres pioneered, requires an accounting of a baseline, any changes, and goal setting. Ultimately, a sustainability report will result, and Ceres provides awards as well as limited kudos even for the most recalcitrant who make incremental changes, with Suncor being the example I've used here.

Certainly, there are others who are working in the same space on CSR issues like Ceres (see Hoffman 2011), but Ceres's tactical work with both investors and companies, producing reports, supporting and leading the charge on shareholder resolutions, and calling for policy changes and regulations set it in a category by itself. What became apparent to me through the course of my research with Ceres is that it tends to think of linguistic and vernacular changes as a kind of product. The uptake of terminology by the business community, writ large, is both the goal and the marker of success. Climate risk is such "a product" in terms of both the effort to conceive of it and the success with which deployment has been met. Unlike the pastors in Creation Care who assume that their words will be paired with others, dissected, and discussed, Ceres's word-products reverberate via marketplace circulation. INCR and the Summit at the United Nations proved to be key mechanisms and testing grounds for furthering the notion of climate risk, and they have attracted leading investors and companies such that the volume reinforces the weight and claim of risk associated with climate change.

Climate risk has come to encapsulate regulatory, litigational, physical, and reputational risks. In order to transmit these risks to both investors and companies, Ceres has built on the infrastructure left in place by anti-apartheid and anti-pollution activism with corporations. Much like apartheid, climate risk has been positioned as both a kind of moral pariah and present danger, where action is required by companies in order to disassociate themselves from it. Unlike apartheid, however, climate risk is fraught with potential unintended consequences suggested through Beck's work in the form of new alliances, assemblages, inequities, and other structural shifts and changes. As well, the notion of risk if we follow Ho's work can undermine arguments for addressing climate change and shareholder value. The higher the risk, as Ho's research narrates, the higher the reward. One only has to consider the continued fallout from the credit crises and subsequent restructuring of the banking industry to find further evidence to support Ho's analysis.

One of the deep concerns for Ceres and its membership has been the reward structure in place through capital markets—more commonly referred to as "Wall Street." Public companies are to a great degree valued by their publicly traded share price on stock markets, which reward with a higher price/valuation based on quarterly earnings. Major changes on emissions reductions usually require investments that will take much longer to see returns than markets have the patience to wait for—hence the importance of a term like *climate risk* as well as the role of INCR and its annual summit. They lobby for disclosure of the amount of risk a company is exposed to related to climatic changes and demand changes to protect their investments, thus giving "cover" to companies who need to make massive changes, infrastructural or otherwise, to address this risk.

In 2010, it looked as though Ceres had seen one of its goals realized when the SEC set out nonbinding guidelines that recommended disclosure on the risk of climate change. SEC commissioners, of which there are five, noted that they were under immense pressure from investors to make this change, yet they were careful to avoid taking a stance on climate change itself. Two of the five commissioners, noted as Republican appointees, voted against this decision, citing the problem that climate change is still "unsettled" in terms of the scientific claims associated with it. Clearly then, not all of Wall Street has come around to the notion of climate risk or the veracity of the scientific claims associated with climate change.

This chapter's account of Ceres provides another facet by which to understand climate change's evolving, emergent form of life. But coming to terms with climate change in this chapter is not about epistemological difference in the classic sense of resisting science or its factual claims. Rather, the risk framework in this instance sidelines debates about the specifics of climate science in favor of the language of business and investment such that action is required both by investors in their assessment of companies and by companies in their strategic planning. Sea level rise and more volatile, disrupted weather patterns provide the impetus for assessing physical risk, while regulation, litigation, and reputational risk are aspects that pertain solely to a business environment. In calling these other considerations into being, providing opportunities for investor and environmental concerns to establish collaborative discursive turns and mechanisms, Ceres establishes itself as a pivotal element of the assemblage of institutions, vernaculars, and articulations.

Epilogue

Rethinking Public Engagement and Collaboration

The critic is not the one who debunks, but the one who assembles. The critic is not the one who lifts the rugs from under the feet of the naïve believers, but the one who offers the participants arenas in which to gather. The critic is not the one who alternates haphazardly between antifetishism and positivism like the drunk iconoclast drawn by Goya, but the one for whom, if something is constructed, then it means it is fragile and thus in great need of care and caution.
Bruno Latour, "Why Has Critique Run Out of Steam?"

On June 25, 2013, President Barack Obama delivered a "new national climate action plan." He began by talking about the Apollo 8 mission in 1968 and the images of the "blue planet" they brought back that transformed perceptions of the earth and humanity. He introduced new targets for carbon emissions as well as changes to domestic regulation and foreign policy. But it was Obama's challenge to the American public that stood out as I put the finishing touches on this book:

What we need in this fight are citizens who will stand up and speak up and compel us to do what this moment demands. Understand, this is not just a job for politicians. So I'm going to need all of you to educate your classmates, your colleagues, your parents, your friends. Tell them what's at stake. Speak up at town halls, church groups, PTA meetings. Push back on misinformation. Speak up for the facts. Broaden the circle of those who are willing to stand up for our future.

Obama's call to Americans to exercise their citizenship by using every avenue available to educate their communities illustrates perhaps most poignantly how the conditions persist that drove the research questions in this book. Yet "speaking up for the facts," as this book demonstrates, is no simple task, not even for those invested in producing and circulating those facts to much wider audiences. It is strongly apparent, however, that persuasion and speaking for the facts involve and enroll social ties and affiliations, but not just in terms of a transfer of information or a "pushback on misinformation." The communal life of facts matters in explicit and implicit ways—it matters in terms of directing attention (and attentional rest), adjudicating debates, deciding what side one wants to be associated with and "what's at stake," and knowing who and what can be trusted. Not only science, but the social determines ethical and moral value—and the consequent rationale to act—helping to resolve challenges to long-held ideals and norms, adapt practices and modes of communicating facts, and navigate epistemological difference.

Climate change requires such a negotiation with ethics, morality, and meaning-making in collective and individual terms—a negotiation that moves us beyond the fuzzily beautiful vision offered by a "blue planet," toward a multifaceted engagement with how facts and information come to matter beyond and within the scientific contexts in which they first emerge. The social and professional groups recorded here bring the multiplicity of such processes, and inherent shifts in norms and practices, into sharper focus. Climate change as a form of life is constituted and defined through its use. Only as this process of collectively defining what climate change means continues to unfold can one ask and begin to answer something akin to Wittgenstein's language games: What does it mean to believe in climate change? What does it mean to have a future with climate change? What will it mean to inhabit that future?

Public Engagement and Experimental Futures

In using the concept of an emergent *form of life* as a method, I've sought to examine the ethical and moral contours by which climate change has come to have meaning collectively and, by extension, individually. I've drawn on ethnographic evidence to construct an experimental system for understanding collective efforts at defining climate change, its rules, grammars, and associations. Each chapter provides multiple points of entry to understanding the ways in which climate change has been translated and rearticulated for and by groups, morphing ideas about who can speak for, about, and to the issue—and *how* to speak about it. Yet this book also provides a tableau for future engagement and an adjoining of knowledges such that negotiation and engagement might occur—even as differences (and resonances) in goals, vernaculars, and epistemologies become apparent.

Climate change poses an intellectual, scientific, and moral challenge. It is a problem of assessing what is happening, what might happen, and how to act in the world, as well as an evaluation of epistemological differences. Who the messenger is, how climate change matters for the group, and how they code it for immediate response and action defines climate change's form of life for that group, but it also generates questions about collaboration when definitions, associations, rules, and grammars differ. What does collaboration mean when goals related to climate risks are differently configured? How much do epistemological differences matter? Configured as differences in epistemology, "speaking up for the facts" might require as much listening as it does speaking.

The presentation and circulation of information provide only partial answers and require a partnership with codes for meaning, ethics, and morality in order to delineate what the stakes and risks entail. Framing long-term uncertain issues in order to generate immediate action requires collaboration. Societies dominated by evolving chains of postindustrial risks must confront the definitions of risks as well as the question of how they will unfold and how to respond (Beck 1992, 2002).

The competition for defining climate change is continually played out in the media, enrolling some translations and rejecting others. Indeed, this is what comprises emergent forms of life (Fischer 2003). Debate in a risk society is over how to define the degree, scale, and urgency of risk and, in so doing, opens up rifts between those who produce and those who consume these definitions (Beck 1992). Media change has begun to create records of these evolutions (and revolutions) through blogging and the recontextualizing and repositioning of mainstream articles in social media.

Verification, professional norms, and accountability will surely continue to be negotiated long after Twitter and Facebook cease to be the dominant platforms for social networking. But what new opportunities for wider public participation have wrought is nothing short of transformational for journalists and particularly so on contentious expertise-laden issues with moral and ethical contours. The role imagined for journalism in our democracy has traditionally been one of informer, agenda-setter, and watchdog, but journalists and news organizations now need to add decidedly different tasks to their obligations—that of forum-provider, chief discussant, and lead verifier. Climate stories are now subject to scrutiny and counterclaims—what Bruno Latour (2004b) has called "instant revisionism"—from concerned audiences with diverse perspectives.

For those engaged in this issue, epistemological concerns are now foregrounded: how evidence is deployed, who is speaking for it, and where scientific knowledge has been produced are vital details. For wider publics, what flows from the "so what" question is a drive to know and understand more, to do something, to adopt a position, to be part of discussions about what *ought* to be done. The ways in which facts are socialized is key to the establishment of climate change if, as is argued by such a wide spectrum of those engaged with the issue, action is required related to the facts.

Throughout this book, I've used the term *public engagement* to signal the need and desire of those invested in educating, informing, and motivating the public to act on climate change. The term *engagement* implies both a desire to find out more about an issue and an ethical obligation to become concerned and to act (Callison 2009). Engagement is not only awareness, nor is it just a matter of getting the facts out. Rather, it is a connection most often visible in our social networks, whether it be our church, ethnic group, political party, workplace, school, or other affiliation. Engagement requires collectivity; it feeds on debate so that ethics and associations with an issue might be established. In part, this is why new media continue to play such a large role in the climate change debates. And I am arguing here that engagement is not something the broadcasting of facts can accomplish on its own.

Not Just the Facts

Even though climate change may have begun as a scientific concept, it has flourished as it's been adopted, torqued, politicized, paired. In short, it's been filled with meaning through its interaction with belief systems, practices, and other systems of knowledge. The discursive strategies investigated here are ultimately heterogeneous, emergent, and multivocal. They defy any "framing" of climate change in a static, solely scientific, or progressive/liberal environmentalist fashion.

Defining climate change as a risk presents a conundrum for how to define its scope. Insofar as science is understood as a spectrum of possibilities, risk framing allows for an accounting for potential benefits and losses. Many scientists perceive the facts and the risk framework as requiring the work of near-advocacy—following scientific expertise as it travels into the public arenas, caring and speaking for the science, and attempting to ensure that it is not being distorted or misused.

Full advocacy would subordinate science to political goals, but near-advocacy follows scientific findings into practical realms while trying to maintain the integrity of professional scientific norms. Some scientists have gone so far as to get involved in policy; others have been pushed by a confluence of factors. Economics is rising as a way to discuss valuation and ethical dilemmas, and since my fieldwork concluded, it has become even more common, particularly as risk assessment becomes a dominant paradigm for understanding climate change. Such assessments submerge epistemic differences, even as concerns arise about exaggeration.

I have theorized that in the instances where science is not the sole evidence upon which decisions are made or positions are struck, scientific findings are partners. It means that at times, in STS terms, the science is "black-boxed," and in others, it is complemented by another knowledge system. The Inuit bring traditional ecological knowledge in the form of oral histories, ground truthing, and other qualitative and quantitative observations and interactions with the natural world alongside science. With Ceres and Creation Care, science is most often put away as settled or, in any case, not up for discussion except in terms of its ethical and/or moral ramifications. Yet science is never completely absent. It hovers in the background, being moved carefully to the foreground when and as needed, however briefly, as an affirmation and to underscore the rationalization or logic already under way.

Journalists have struggled at the local level with the ways in which climate change is either empty for large swaths of the public, seen largely as a remote futuristic scientific concept still being debated, or full in the politically partisan sense. And while these struggles go on at the national level as well, the negotiation is quite different. Where the science is settled, the immediacy and implications of it are not. Figures like Andrew Revkin and Boyce Rensberger adjudicate expertise, advocacy, and near-advocacy in an attempt to fulfill the role of arbiter for the public, in their role as a fourth estate, watchdog, worthy of the public trust. Journalists' expectations are that others will trust them while they trust no one, not even the experts. Cornelia Dean explained it to me as an old adage in journalism: "If your mother says she loves you, you should check it out." As blogging and forms of social media proliferate, it has become increasingly apparent that journalists are being checked out and are increasingly required to verify others as well (Kovach and Rosenstiel 2007).

This question of expertise and what form of climate change it represents is one this book foregrounds. It is in some ways a continuation of debates about the role of expertise in the public domain, one begun by Walter Lippmann in the 1920s. However, unlike Lippmann's formulation, the question of expertise is not only about dominant knowledge paradigms. Instead, the social groups I have studied here ask who gets to speak for climate change and how it might be defined in their terms. Expertise in this sense is being morphed by those who are investing climate change with particularities all the while reinforcing its universality and status as a multifaceted form of life.

Climate Change as Risk and Reminder

Risk opens up ethical challenges that can't be understood only as problems of expertise or translation. Kim Fortun's analysis of the aftermath and advocacy after the Bhopal gas explosion led her to ask how to account for the ways in which disaster creates community. She theorized that enunciatory communities come together in response to a temporary paradox, as a result of contradiction, force, and double binds. Advocates establish "how the past should be encountered" and "what counts as adequate" in terms of description and explanation. Fortun is describing something different—climate change is a *slow motion disaster* by contrast, but there is a similarity to the global connectedness, exclusions, expansiveness, ethics, and predictions of catastrophe she describes.

Reading climate change through Fortun's work led me to ask: how do we account for the way risk creates or, rather, reminds us that we are community? Climate change presents a range of predictions that vary from mild and inconvenient to world-altering, even near-disaster-movie kinds of scenarios, if we include abrupt climate change within the range of possibilities. It makes clear that what gets put up into the atmosphere circulates and has an effect on polar communities far from the origins of most greenhouse gas emissions (notwithstanding the north's own grid related to housing and transportation). For those with a security focus, it makes clear how dependence on natural resources can exacerbate seeds of conflict that might, in prosperous times, go uncultivated. In short, it breaks down many of the barriers that wealth and power have erected between geographically and socially disparate places. And climate change provides the impetus for creating networks fueled by much of the same wealth and power through routes established by science, media, and national and international policy.

The issues of justice, equity, and connectedness sit uncomfortably on the terrain of mitigation and adaptation solutions—of who and what are considered vulnerable enough to warrant immediate action. As Stephen Schneider put it to the group of reporters in Oregon, the debates are most often about *fairness*. Advocacy on climate change attempts to establish how the future should be encountered and considered. This is most evident in policy discussions and economists' debates about the discount rate. But policy ties all the groups together. It is the intervention that may, by many estimations, determine what the future will look like. Attempts at intervening into or affecting policy make evident the moral and ethical codes and norms of each group.

Creation Care's interventions into policy about climate change have been about concern for impoverished countries (albeit an undefined, somewhat utopian vision of otherness and poverty) in American legislation. The Inuit have intervened regarding cultural and communal survival at the international levels with the human rights petition, seeking to put pressure on the United States. They continue to work at the international, regional, and national levels of policymaking so that their voice is heard within the Arctic and as an Arctic voice reminding the world that there are people and cultures at stake at one of the poles. Ceres has been working on legislation that might put a price on carbon and regulate emissions, but also at the level of the SEC so that disclosure of risk might be regulated. Science experts work with the IPCC or through other mechanisms in order to see the science predictions receive the response they warrant—the work I've described as near-advocacy.

Predilections toward alarmism (justified and otherwise) are very much related to how one experiences a future with climate change and what ethics one applies to the portent of such a vision. One journalist I heard speak said it this way: alarmism is needlessly ringing the alarm bell, but what if the alarm bell needs to be rung? Sheila Watt-Cloutier put it even more strongly when she said, "I think that some people have not fully come to understand that there is no disconnect between suicide and climate change." In other words, social problems, the continuance of culture, and the state of the Arctic are bound together. In this formulation, climate change continues the process of foreclosure on hope, begun by encounters with colonialism and the enduring structures it put in place via education and the now slowly evolving mechanisms for governance and self-determination.

What this brings to the foreground then is *the route* between feeling, experiencing, knowing climate change as either a prediction or lived experience and making changes to policies that might address the causes and effects of climate change. That route is multifarious and marked by shifting assemblages and institutions called into being by the heralding of climate-related risks. The social groups thus provide an added dimension of those outside the juggernaut of news-making scientists and formal policy negotiations. Bringing them together exposes the ways in which vernacular guides the formulation of climate change as an experience, dictum, and ethical directive both for the group and the public at large. Climate change becomes the starting point for explaining how we, as global communal members, fit together under a rubric of ethically and morally shaped relations. But that "fit" is a moving target and one that plays differently depending on what nationalistic, professional, or capital-oriented audience climate change is being presented. This is indeed what makes it difficult to engage wide publics with climate change as a fact requiring action.

As I pointed out in chapter 5, using Beck's formulation, risk acts to unite societies, but it also creates new loci of conflict, alliances, inequalities, exploitation, and regionalizations. At the geopolitical level, this is what much of the climate negotiations at the UNFCCC or IPCC make wholly evident. But policy negotiations provide only one part of the equation. The shifting assemblages of media, bureaucracy, institutions, and advocacy groups provide another window on the ways in which risk both acts as a herald of change and calls it into being. Climate change effects change precisely because of its status as a risk that could reorient topographies of wealth and power. How or whether the status quo can be maintained or disrupted depends on the view one has of both the present and the future.

Justifying and calling for such changes require a momentum that begins for many with shifts in public opinion, and for this, media and communication in general remain crucial. This in part explains the calls for the media to do better at getting the point across, motivating the public, getting the science right so that the fact that something should be done becomes self-evident. Yet has media been expected to do what it is incapable of doing—investing meaning, ethics, and morality in a particular issue such that the public is called to act?

In thinking about the role of media and information, I have built on observations by Herbert Gans and Michael Schudson that information does not necessarily lead to participation. And yet American democracy, to a large extent, is built on the notion that we need to be informed in order to participate, in order to do our duty as citizens, whether it be casting a vote or advocating for change. That information is seen to flow largely through media.

The rise of new media has made apparent the wide variety of voices and responses to mainstream discourse. I had expected to find the social groups I researched to be heavy users of the renegade change application that is new media, but many were still focused on getting and keeping the attention of mainstream and/or leading reporters like Andrew Revkin or, as was the case with Ceres, getting the attention of the business press. This was beginning to change toward the end of my research as the full potential of social media began to emerge. The confluence and difficulty of strategizing with and for media, as is often the case with social groups with limited resources, illustrate just how tenuous and changeable the media landscape is.

Andrew Revkin, about five months into running his blog *Dot Earth*, came to an event I helped organize at MIT in 2008.[1] During his panel presentation, he summarized a conversation he had with the late Stephen Schneider. Revkin had told Schneider that the questions related to confronting sustainability issues were much broader than journalism alone. Revkin asked: "Can journalism handle it, and can science handle it?" To which Schneider replied: "The question is whether democracy can survive complexity." Certainly, when information is at the heart of such a question, it gets more and more difficult to see how the public and policy-making can adjudicate increasingly complex interests, information, and resulting configurations of possible action. Climate change does indeed present such a challenge to democracy—one that requires a shift toward acknowledging the challenge of defining emergent forms of life and negotiating with varied ethics, morality, and meanings.

Appendix:
A Decade of Climate Change

Since the release of the 3rd Assessment Reports from the Intergovernmental Panel on Climate Change (IPCC) in 2001, there have been many efforts to engage the American public. This is not a comprehensive timeline but an overview of events and changes relevant to this book.

2001 The IPCC, an international organization of thousands of climate scientists, representing 130 countries, and formed by the World Meteorological Organization and United Nations Environment Programme, releases its 3rd Assessment Reports.

— Sheila Watt-Cloutier begins four-year term as chair of the Inuit Circumpolar Council (ICC).

— Following the founding of their Energy and Climate Change program a year earlier, Ceres, a corporate social responsibility group based in Boston, meets to discuss how to link risk to climate change.

— Gallup begins polling Americans about whether news coverage of climate change is exaggerated, correct, or underreported. By 2010, those who see news coverage as exaggerated will reach a high of 48 percent.

— Newly elected president George W. Bush rejects the United Nations Framework Convention on Climate Change (UNFCCC) Kyoto Protocol.

2002 The Evangelical Environment Network launches its "What Would Jesus Drive?" campaign.

— John Houghton, chair of IPCC's 2001 Working Group One, joins with American scientist Calvin DeWitt to begin a distinctly evangelical dialogue on climate change. Together, their two groups, the John Ray Initiative (Houghton) and Au Sable Institute (DeWitt), organize a conference for Christians at Oxford. Richard Cizik, the former vice president of government affairs for the National Evangelical Association (NAE), is a reluctant attendant, but later describes his experience as one of "conversion" to concern about climate change.

2003 Republican strategist Frank Luntz, in a memo to President Bush that was later leaked, deemed "climate change" a less threatening term than "global warming" and advocated its use by the Bush administration. The memo so influenced environmentalists that, for a time, many opted for "global warming" (Lee 2003).

— Ceres holds its first biennial Investor Summit on Climate Risk at the United Nations in New York City. By 2012, INCR will have 520 participants representing assets totaling $22 trillion.

2004 The Arctic Council releases the Arctic Climate Impact Assessment (ACIA). Billed as a thorough combination of traditional and scientific knowledge, ACIA heavily involved indigenous people and over 300 scientists.

— The Union of Concerned Scientists issues two reports and a statement signed by sixty prominent scientists, including twenty Nobel laureates, condemning the Bush administration's "distortion" and "misuse" of science and in particular castigating the inaction on climate change, which scientists had nearly unanimously agreed should be a priority issue.

— The intention to draft an Inuit human rights petition is announced at the annual UNFCCC Conference of the Parties (COP) 10 in Buenos Aires. The petition is led by then international ICC chair and Canadian Inuk Sheila Watt-Cloutier.

— *Science* publishes an article by historian of science and geologist Naomi Oreskes, who analyzed 928 peer-reviewed scientific articles and argues that scientists are in consensus on climate change. She states: "Politicians, economists, journalists, and others may have the impression of confusion, disagreement, or discord among climate scientists, but that impression is incorrect." Oreskes subsequently writes an op-ed based on this research for the *Washington Post* (2004a; 2004c).

— Cizik, Ball, and DeWitt attempt to re-create the Oxford experience in the United States at another conference held near the headwaters of the Chesapeake Bay at the Sandy Cove Conference Center. Houghton is a keynote speaker, and the conference is sponsored by the Evangelical Environment Network, NAE, and *Christianity Today*. The conference produces the Sandy Cove Covenant and lays the foundation for the 2006 Evangelical Climate Initiative (ECI).

— NAE releases "For the Health of the Nation: An Evangelical Call to Civic Responsibility," in which it lists Creation Care as one of its priorities.

— An article by Maxwell and Jules Boykoff is published in *Global Environmental Change* showing that media are biased in their reporting of climate change because they have reported equally on climate skeptics, adhering to the journalistic practice of balancing opposing points of view.

2005   Signed in 1997, the UNFCCC Kyoto Protocol comes into effect.

— Michael Shellenberger and Ted Nordhaus publish their essay, "The Death of Environmentalism," contending that the public's inability to attend to climate change in any significant way is prime evidence of the environmental movement's failure. This sets off a round of intense debates and discussions within the environmental movement, including a response from environmental justice advocates called "The Soul of Environmentalism."

— Hurricane Katrina formed over the Bahamas, hit the coast of Florida, traveled up the Gulf of Mexico, and made landfall as a Category 5 Hurricane on August 23, wreaking destruction over 100 miles inland throughout the Mississippi region—most severely on the city of New Orleans. Though the exact number remains in dispute, it is estimated that almost 2,000 people died. It was the costliest natural disaster in U.S. history. Shortly after Katrina, Hurricane Rita followed, adding yet more injury and loss to an already battered Gulf coastline.

— MIT's Kerry Emanuel publishes an article in *Nature* that finds a link between climate change and a rise in the intensity of hurricanes, but his projections are for fifty years hence. Still, he becomes a media sensation and is inundated with calls from reporters in the days following the storms. Based in large part on Emanuel's research, *Time* magazine devotes a cover the week after Hurricanes Katrina and Rita asking: "Are We Making Hurricanes Worse? The Impact of Global Warming" (Kluger 2005). Emanuel was later profiled in the *New York Times* and other major newspapers, and *Time* named him one of the most influential people in 2006 (Dreifus 2006; Kluger 2006). It didn't matter that Hurricane Katrina could not be directly attributed to climate change, nor could the damage it inflicted be solely attributed to the hurricane itself. Investigations afterward found that the failure of levees in New Orleans, a foreseeable technological problem, led to much of the worst damage (McQuaid and Schleifstein 2006).

— On December 7 at the UNFCCC COP 11 in Montreal, a group gathers for a side table session called "the right to be cold." Sheila Watt-Cloutier announces that, after two years of research, she and sixty-two other Inuit individuals have submitted a petition to the Inter-American Commission on Human Rights. The petition named the United States as a violator of the 1948 Declaration of the Rights and Duties of Man. The petition states that U.S. inaction on reducing greenhouse gas emissions to mitigate the effects of climate change violates the Inuit right to life and physical security, personal property, health, practice of culture, use of land traditionally used and occupied, and the means of subsistence.

2006 In May 2006 *An Inconvenient Truth* premieres, featuring Al Gore and his climate "slide show," arguing for the scientific fact of climate change and the need for personal and political action. It features prominently the imagery from and following Hurricane Katrina, as well as other evidence of what climate change could portend for humanity on a global scale. The film later receives an Oscar in the documentary category.

— The Evangelical Climate Initiative, a declaration signed by a group that includes mega-church pastors, Christian college presidents, and para-church leaders, is released. The *New York Times* reports in advance of the release of ECI that other prominent evangelical leaders including James Dobson and Charles Colson have sent a letter declaring that ECI does not speak for all evangelicals. This letter spawns the Interfaith Stewardship Alliance and reinvigorates the 1999 Cornwall Declaration, which maintains that the science behind climate change is still "uncertain."

— Sir Nicholas Stern releases *The Economics of Climate Change: The Stern Review* in November 2006 and unleashes a maelstrom of controversy in science, policy, and media circles. Commissioned by then UK prime minister Tony Blair, Stern's 700-page report focuses on whether it makes sense or not to pursue a mitigation strategy in the face of a range of scientific climate change predictions. Its main conclusion is that the benefits of strong, early action on climate change far outweigh the costs of not acting, and that "climate change will affect the basic elements of life for people around the world. . . . Hundreds of millions of people could suffer hunger, water shortages and coastal flooding as the world warms." Stern estimates that not acting to avert climate change could cost the equivalent of losing 5 percent global GDP annually—a figure that could rise to 20 percent "if a wide range of risks and impacts is taken into account." Reducing greenhouse gas emissions now, in contrast, could limit those costs to 1 percent annually.

— The Inuit petition to the Inter-American Commission on Human Rights is rejected. Sheila Watt-Cloutier is later asked to make a submission in early 2007 on behalf of all indigenous peoples adversely affected by climate change.

2007 The IPCC releases its fourth series of assessment reports. For the first time, the IPCC devises a media strategy complete with press conferences in order to avoid the kind of lackluster response they had gotten in 2001. Taking off the scientific qualifications found in previous reports, the IPCC states that the warming of the climate is unequivocal and the global temperature increase is very likely due to anthropogenic greenhouse gas emissions. The report predicts that warming, Arctic ice melt, and sea level rise would continue for decades even if emissions levels were stabilized and that the attendant impacts could be catastrophic for island nation-states and those dependent on polar ice caps.

— On January 17, 2007, a group of twenty-eight scientific and evangelical leaders issue "An Urgent Call to Action: Scientists and Evangelicals Unite to Protect Creation." The "Call" speaks convincingly about the shared "moral passion" and "sense of vocation to save the imperiled living world before our damages to it remake it as another kind of planet." It states that the protection of life on earth "requires a new moral awakening to a compelling demand, clearly articulated in Scripture and supported by science," and it specifically expresses concern for "the poorest of the poor" who not incidentally also inhabit some of the richest areas of Earth's biodiversity.

— Arctic Science Summit, held at Dartmouth College, kicks off the fourth International Polar Year (2007–9). IPY is sponsored by the International Council for Science (ICSU) and the World Meteorological Organization (WMO), and is backed by national funding bodies like the National Science Foundation in the United States. The first IPY was held in 1882–83, and included eleven participant countries, and fifteen Polar expeditions. The second IPY was held 1932–33 and included forty participant countries. The third IPY was held in 1957–58 in conjunction with the International Geophysical Year—an event proposed by the International Council of Scientific Unions. Sixty-seven nations participated, with twelve nations participating through sixty-five research stations in Antarctica.

— The Supreme Court rules that the U.S. Environmental Protection Agency (EPA) is legally required to account for greenhouse gas emissions.

— President Bush acknowledges the veracity of climate change in his statement that Americans are "addicted to oil." Despite this shift, six climate-related bills in the Senate, and eight in the House of Representatives, no formal changes in American policy ensue.

— Al Gore and the IPCC are awarded the Nobel Peace Prize "for their efforts to build up and disseminate greater knowledge about man-made climate change, and to lay the foundations for the measures that are needed to counteract such change." Earlier, Sheila Watt-Cloutier had been rumored to be a potential recipient of the prize.

— Al Gore joins with Kevin Wall to produce Live Earth on July 7, 2007 (7–7–7), a series of benefit concerts occurring in Sydney, Johannesburg, Rio de Janeiro, Tokyo, Kyoto, Shanghai, London, Hamburg, Rome, Washington, D.C., East Rutherford, New Jersey, and Antarctica. The concerts involved 150 artists and were broadcast in 132 countries.

— Media coverage of climate change in the United States hits an all-time high.

2008 Creation Care holds its first conference to equip pastors at Northland Church in Orlando, Florida.

— Polar bears are made an endangered species as a result of climate change predictions, but many critics who reside across the Arctic in Canada and the United States decried this as a symbolic and ultimately unhelpful change to policy (Economist 2007; Palin 2008; Watt-Cloutier 2007).

— With one of the longest running records of public opinion on the issue, Gallup reports in April that climate change continues to rank near the bottom of environmental concerns and that those who worry about climate change "a great deal" are about the same percentage as in 1989 when they first began polling on the issue (Newport).

— A 2008 survey by the Pew Research Center finds that evangelicals (31 percent) are more likely than the average American (21 percent) and much more likely than mainline Protestants (18 percent) and black Protestants (15 percent) to deny the existence of climate change and anthropogenic causes. And while 47 percent of Americans acknowledged there was "solid evidence" of climate change and human causality, only 34 percent of white evangelicals and 39 percent of black evangelicals agreed. These percentages are lower than the percentage of Republicans in general who are not convinced of the fact of climate change. During this same period, the percentage of Republicans convinced of climate change began to decrease from 62 percent in 2007 to 49 percent in 2008 as compared with 84 percent of Democrats and 75 percent of independents (Pew 2008).

— Barack Obama is elected president for the first time.

2009 A Global Summit of Indigenous Peoples is held in Anchorage, Alaska, to discuss climate change, led in part by ICC chair Patricia Cochran, now two years into her four-year term following Watt-Cloutier. The resulting Anchorage Declaration is meant to provide direction to the upcoming UNFCCC COP 15, but not all parties present at the Summit agree to sign on.

— Several weeks before COP 15, e-mail accounts belonging to the Climatic Research Group at the University of East Anglia are hacked. Scientists in the United States and Great Britain are accused of bias in the peer review discussions and of manipulating or misrepresenting data. Multiple review committees later exonerate the scientists involved, finding that peer review had not been compromised.

— UNFCCC holds its COP 15 in Copenhagen. According to COP 13, held in Bali in 2007, a new framework for climate change mitigation should be adopted at Copenhagen. The resulting Copenhagen Accord was roundly criticized as a failure to achieve binding targets for emissions reduction.

2010 The SEC releases nonbinding guidelines that recommend climate change risk disclosure. Months later, however, the Chicago Climate Exchange, a voluntary, legally binding emissions trading exchange and the only one of its kind in the United States with 450 members, ceases trading. The European Climate Exchange and Chicago Climate Futures Exchange remain operational, but the loss of CCX deals a major blow to many who touted it as a market-based solution in advance of policy.

— In response to what is widely viewed as the failure of COP 15 in Copenhagen, the Bolivian government hosts the World People's Conference on Climate Change and the Rights of Mother Earth in Cochabamba, Bolivia, resulting in a declaration entitled "the Peoples Agreement."

2012 The first commitment period for emissions reduction related to the Kyoto Protocol (covering 2008–12) comes to an end.

2013 Following Hurricane Sandy, President Obama says this in his State of the Union address on February 12: "But for the sake of our children and our future, we must do more to combat climate change. Yes, it's true that no single event makes a trend. But the fact is, the twelve hottest years on record have all come in the last fifteen. Heat waves, droughts, wildfires, and floods—all are now more frequent and intense. We can choose to believe that Superstorm Sandy, and the most severe drought in decades, and the worst wildfires some states have ever seen were all just a freak coincidence. Or we can choose to believe in the overwhelming judgment of science—and act before it's too late." Obama announces his national climate action plan on June 25.

— As this book goes to press, global atmospheric concentrations of carbon dioxide have hit 400 parts per million for the first time in 3 million years, and the IPCC is releasing its fifth assessment reports beginning in late 2013. Many hold high hopes for COP 20 in Lima, scheduled for December 2014. Gallup polls indicate that 58 percent of Americans are concerned about climate change, up from 51 percent in 2011. And while 48 percent believed media coverage of climate change was exaggerated in 2010, only 41 percent believe the same in 2013.

Notes

Introduction

1. By double bind, I am drawing on Kim Fortun's definition in *Advocacy after Bhopal* where she analyzes situations in which "individuals are confronted with dual or multiple obligations that are related and equally valued, but incongruent" (2001, 13). Double binds cannot be resolved or reduced through narratives; they present persistently mismatched messages and explanations that must be accounted for simultaneously.

2. This book project started out as a dissertation project that has its earliest roots in my first experience with debates about climate change and the American public in the fall of 2002, sitting in a class at the Massachusetts Institute of Technology, where I attended graduate school. The class focused on global environmental problems with topics ranging from climate change to nuclear power, coal production, and industrial pollutants. My classmates were all young scientists and engineers conducting graduate research on environmental problems. On the day that climate change took center stage, we moved beyond the usual discussion of science and policy. To my surprise, many students angrily blamed the media for not getting the public to care about climate change as an issue that required both personal and collective action. The solution, some students forcefully argued, was that we needed an image to capture the problem and galvanize public imagination in the same way that the ozone hole had, for example. Others argued that what we needed was an easy, direct action response like a blue recycling bin to set out beside the garbage can. I responded by arguing for a complex view of media and its changing role in society. A few others valiantly tried to intervene, contending that much-needed solutions were not easy to implement or visualize. But our voices were largely drowned out as the class discussion concluded at a crescendo of frustration with media for not doing their job—that of informing *and* convincing the American people to care about climate change. Massive shifts in media generally and journalism specifically have become that much more evident in the decade since this initial class discussion. Yet arguments and assertions that view major media in the United States as an agenda setting, objective fact dispensing institution that can and should move democratic publics and their polities continue to persist and hold sway far beyond the classroom.

3. Sheila Jasanoff (2004) defines coproduction as an idiom that acknowledges the way scientific knowledge both embeds and is embedded in social practices, identities, norms, institutions, and discourses.

4. Semistructured interviews with leaders and participants were then conducted during and after such collaborative spaces in order to more fully understand or clarify the discursive processes at work. Historical and current publications and scholarly literature provided a secondary site for analysis. In addition, a para-site was constructed by way of "Disruptive Environments: Activists and Academics in Conversation," a student-organized conference at MIT in April 2008 in which a panel I organized on climate change featured several of the prominent interviewees for this book in conversation with one another.

5. Fortun formulated the idea of enunciatory communities in order to account for the way disaster creates community (2001). It differs significantly from the communities followed in this research because the Bhopal advocacy networks grew out of a specific incident and were strategically and temporarily configured. The groups I look at are long established and are reconfiguring their messages and positions in the face of a new global and present/future crisis. In other words, they did not emerge out of crisis like the groups that Fortun studies. Yet there is a resemblance to the ones Fortun describes particularly in relation to the multiple identities many groups and group leaders profess and the sense of divided loyalties while attempting to intervene amid a sea of cultural commentaries. Hence the paradox of double binds, geographic dispersal, epistemological inconsistencies, and multivocality that she describes echo throughout my research.

6. I thank an anonymous peer-reviewer who later revealed himself to me as Timothy Choy for his tremendously helpful insight on this point.

7. See Bloor 1983.

8. See Gayatri Spivak's interview in the *Post-Colonial Critic* (1990).

9. The Sapir-Whorf thesis was based on their research with Indigenous American groups and posits that a group's language helps to form their worldview: "The categories and types that we isolate from the world of phenomena we do not find there because they stare every observer in the face. On the contrary the world is pre-

sented in a kaleidoscopic flux of impressions that have to be organized in our minds. This means, largely, by the linguistic system in our minds" (Whorf 1956, 212, and quoted in Kay and Kempton 1984). Linguists had discarded this hypothesis until recently when it has been taken up most prominently by Stanford's Lera Boroditsky (2001), who found that English and Mandarin speakers think about time differently due to their language structures. I am positing here that because climate change is caught up in underlying views about nature, the role of humanity in it, and concepts of the future, these groups also have a distinct way of processing these kinds of issues. How they talk about climate change reveals their worldview; the language they use to motivate their group about this abstract scientific concept stems from their group's views of nature and embedded ethical and moral imperatives.

10. See M. Boykoff and A. Nacu-Schmidt, 2013. World Newspaper Coverage of Climate Change or Global Warming, 2004–2013. Cooperative Institute for Research in Environmental Sciences, Center for Science and Technology Policy Research, University of Colorado, last accessed 01 September 2013, http://sciencepolicy.colorado .edu/media_coverage. This page is updated on a monthly basis.

11. For a range of examples from various interdisciplinary academic and nonacademic perspectives, see DiMento and Doughman 2007; Gelbspan 2004; Gore 2006; Hoggan and Littlemore 2009; Lahsen 2005b; Latour 2004b; Leiserowitz et al. 2008; Miller and Edwards 2001; Moser and Dilling 2007; Nisbet 2009; Nisbet and Mooney 2007; Nisbet and Scheufele 2009; Oreskes 2004a; Oreskes and Conway 2010; Shellenberger and Nordhaus 2005; Speth 2004; Ungar 1992; Ward 2008.

12. Though this is the dominant paradigm currently, the Bernal-Polanyi debates of the 1930s represent one prominent moment when the role of science, whom it should serve, and its independence were called into question. See Rouse 1992 for a condensed summary of the debates and how they fit into science and technology studies (STS) thought. A second prominent moment occurred in the aftermath of the successful creation of the atomic bomb. Many scientists including those involved with the Manhattan Project became involved in arms control and antinuclear activism to one degree or another. The Union of Concerned Scientists (UCS) and Pugwash Conferences on Science and World Affairs grow out of such turns to advocacy and activism. UCS continues to be heavily involved in climate change as its veracity and scientists' independence and ability to speak freely come under attack.

13. Nisbet and Scheufele (2009) argue that public engagement is not necessarily about getting more or better information but rather about scientists focusing on communicating science more strategically in order to be more accessible and relevant to Americans. My argument extends much further by investigating how advocacy functions for scientists invested in communicating climate change to wide audiences and by narrating the work of those who bring their own codes of ethics and morality alongside scientific facts in order to produce a rationale for personal, group, and political/policy action. See also Boykoff 2011 and Hulme 2009.

14. In her 2003 analysis of Rayna Rapp's work among women in the 1980s who were grappling with new amniocentesis procedures and technologies, Fortun defines scientific literacy not as "being able to differentiate truth from falsehood, but

about being able to draw on a wide array of discursive resources to understand and make judgments about technoscientific phenomena." In other words, scientific literacy is as much about the process of meaning-making as it is about the facts themselves.

15. Habermas's theory of a public sphere (1962) has been the subject of long sustained scholarly debate. Schudson, in particular, claims that such a rational space for deliberation has never existed. But certainly there is cause for some concern, which Gitlin (2002) in particular articulates regarding the disappearing space for broad public discourse that brings together divergent points of view—something Anderson (1991) calls an "imagined community." Cable television and so-called affirmative journalism, which reinforce certain points of view, are also part of this concern. Jenkins (2006a), Castells (2009), and Benkler (2006) point out that these wide public spaces, while fragmented, have moved online in the form of wikis, blogs, and other networking devices. But as Dean argues in Boler 2008, the problem of too many messages creates an accountability vacuum, which she terms "communicative capitalism."

16. The Project for Excellence in Journalism, which produces "The State of the News Media" report annually, called the transformation facing journalists "epochal" and stated unequivocally in their 2007 report that "technology is redefining the role of the citizen—endowing the individual with more responsibility and command over how he or she consumes information—and that new role is only beginning to be understood. Each annual report since (through to 2013) continues to record the conundrum that blogging, citizen and participatory journalism, and now social media present in terms of ethics, audience, and economic models.

17. In the late 1990s, Bill Kovach and Tim Rosenstiel recognized early indicators of these changes as being part of an ongoing crisis of public trust in journalism. They convened serial meetings of the Committee of Concerned Journalists, and later wrote a book about these meetings that serves as a kind of textbook for what journalists should consider in the pursuance of "storytelling with a purpose." The 2007 edition deals explicitly with the rise of Internet and networked technology, and they identify both a vital ongoing need for professional verification and an elevation of dialogue that is inherently missing in online public forums such as blogs and message boards. Verification in particular raises the need for and methods by which expertise on any given issue is sought. It also speaks to the ways in which citizens are increasingly seeking new and varied information sources as opposed to the more standardized sources of the broadcast era.

18. For more on professionalization related to Lippmann's ideals, see Schudson's *Discovering the News* (1978) and Gans's *Deciding What's News* (1979). For the most recent discussions related to professional norms and practices, see the edited volumes *Rethinking Journalism* (2013) and *The New Ethics of Journalism* (2013).

19. See, e.g., Wohlforth 2004.

Chapter One:
The Inuit Gift

1. Cochran resigned her post as chair in 2009, one year before Alaska's four-year term as international chair ended and after my fieldwork had concluded. James Stotts from Barrow, Alaska, took her place. In 2010, in Nuuk at the ICC Assembly, Aqqaluk Lynge became the new ICC international chair and chair of ICC Greenland. A new chair will be elected as this book goes to press.

2. I use *Inuit* for consistency with ICC, but ICC Alaska represents Iñupiat, Yupik, and Cupik. ICC was formed in the mid-1970s through the initial efforts of Barrow mayor Eben Hopson. Antecedents can be traced back as well to two 1973 conferences held in France and then Greenland that sought to bring together northern indigenous peoples from the Arctic and sub-Arctic. See iccalaska.org.

3. Similar research undertaken in southeast Alaska by University of Alaska Fairbanks doctoral candidate Elizabeth Marino and Professor Peter Schweitzer reveals the same. And when presenting this research at a conference, I was approached by an Australian woman who said she encountered the same emerging problem among Australian farmers who didn't recognize climate change in media as the discursive, experiential object they knew so well. In neither of these cases does it have anything to do with education or ignorance of science. The term held little meaning or connection for them, though they certainly understood what the term was attempting to convey.

4. As David Bloor has pointed out, Wittgenstein brought the study of language into the realm of the social and cultural by treating "cognition as something that is social in its essence. For him [Wittgenstein], our interactions with one another and our participation in a social group were no mere contingencies. They were not the accidental circumstances that attend our knowing; they were constitutive of all that we can ever claim by way of knowledge" (1983, 2).

5. There are significant similarities and differences in each national context. This chapter can't elaborate on these contexts fully. Here is a brief but incomplete index of historical and anthropological work that has informed my sense of the historical contexts in which ICC operates. For Canada, see Alia 2007; Brody 1991; Dahl et al. 2000; Inuit Tapirisat of Canada 1977; Loukacheva 2007; Mitchell 1996b; Semeniuk 2007; Simon 1996. For Alaska, see Anders and Langdon 1989; Arnold 1978; Daley and James 2004; Mason 2002; Mitchell 1997. For Greenland, see Malaurie 2007; Loukacheva 2007; Lynge 1993. For Russia, see Achirgina-Arsiak 1992; Fenge 1999; Vakhtin 1994; Xanthaki 2004.

6. Alvanna-Stemple later resigned as chair when she took a job in Alaska Senator Lisa Murkowski's office in Washington, D.C.

7. Canada has since pulled out of the Kyoto Protocol, and its conservative government had taken a markedly different stance toward climate change at the time that this book is going to press.

8. In fall 2006, NPR's *Marketplace* did a series called "Frozen Assets," in which they looked at new business opportunities or challenges stemming from climate change. The series includes stories on breweries benefiting from melting glaciers in Greenland, potential railroad barons in Churchill, Manitoba, and the Inuit position on offshore oil and gas exploration in Barrow, Alaska. In the story on Barrow, Richard Glenn, vice president of the Arctic Slope Regional Corporation, began his interview with an NPR reporter by saying, "We're not the canaries in the coal mine. It renders you speechless to even toss out a sentence like that." The reporter describes himself as somewhat taken aback by the aggressive opening salvo wherein Glenn is preempting the usual depictions of Inuit. Glenn was closely profiled in Wohlforth 2004.

9. Cruikshank (2005) argues: "Scientists look for physical mechanisms. Oral tradition bearers more often look for moral relationships. Sometimes the narratives return us to a time long ago when giant animals competed with humans for control of the world; in these stories glaciers are the dens of giant animals, and they surge when the animal is angered by thoughtless human behavior. Together the two different approaches give us a richer sense of landscape than can be derived from either one alone" (33). Such fantastic mythological stories likely would seem incongruous to the careful measurements, hypothesis, and analysis that a scientist undertakes. It takes a careful ear and often an exhaustive comparative analysis of the geologic and historic record to reconcile narratives with events.

10. See Harley 1988.

11. The first IPY was held in 1882–83 and included eleven participant countries and fifteen Polar expeditions. The second IPY was held in 1932–33 and included forty participant countries. The third IPY was held in 1957–58 in conjunction with the International Geophysical Year—an event proposed by the International Council of Scientific Unions. Sixty-seven nations participated, with twelve nations participating through sixty-five research stations in Antarctica. See http://www.us-ipy.org for more.

12. The exhibit itself emphasized both the diversity and similarity of Inuit customs and beliefs across a wide geographical area. In other words, Nicole was taking the baby myth from her fieldwork, but that doesn't necessarily mean it is a shared belief across the Arctic, where Inuit dialect and custom vary.

13. Igor Krupnik, the co-curator of the Smithsonian exhibit, was also on my ASSW tour. He warmly complimented Stuckenberger on the *Thin Ice* exhibit.

14. Wohlforth's 2004 work on Alaskan views of climate change and anthropologist Hugh Brody's 1997 work on mental maps and hunting in northern British Columbia certainly come to mind in thinking through this.

15. This text was later reprinted in the exhibit catalogue, which was distributed at the ASSW.

16. See Wohlforth 2004 for a deep description of this relationship and the Barrow Arctic Science Consortium.

17. In the Canadian context, there is a striking comparison to be made here between this petition and the *Delgamuukw v. British Columbia* Supreme Court case, begun in 1984 and concluded in 1997. *Delgamuukw* mandated a revisiting of unextinguished aboriginal titles and rights, particularly in British Columbia, where no trea-

ties with aboriginal people were made before or after Confederation. Wet'suwet'en and Gitksan elders (of which Delgamuukw is the name of one) similarly describe their testimonies and depositions metaphorically as opening up their culture, providing a kind of "gift" to the courts. See Daly 2005; Mills 1994; Monet and Wilson 1992. Such presentations make indigenous ways and practices legible for nontribal audiences, ultimately, so that these audiences (courts and the wider public) can adjudicate the veracity of their claims and, one could extrapolate, the integrity of their culture in relation to the lands they claim. Niezen (2003) makes a similar point with regard specifically to the deployment of a human rights framework by several transnational indigenous organizations.

18. In some ways, this may have helped to pave the way for the work Cochran undertook in her role as ICC chair with the Indigenous People's Summit on Climate Change held in Anchorage. This occurred the year following my fieldwork in 2009. I watched via live webcast. Ultimately, several factors prevented the Anchorage Declaration from having the kind of impact originally intended. It was difficult to ascertain what was going on via remote viewing. Later the World People's Conference on Climate Change in Cochabamba, Bolivia, held in April 2010, produced the People's Agreement of Cochabamba, which seems to have both supplanted the Anchorage Declaration and built upon the groundwork that the summit laid. See Lindisfarne 2010 for more analysis of the Cochabamba event.

19. ICC comprises national organizations and the international chair position, whose office is generally supported by the national office of the current chair's country. The international chair rotates every four years between countries with the exception of Russia, where there is not the infrastructure required to accommodate it. Each country puts forward its own chair and a vice chair to make up the nine-member executive council, which includes the international chair. They meet twice a year as a council. The seventy-two delegates (eighteen from each country) who elect the international chair are chosen differently, depending on the country. Election processes for national chair and vice chair also differ. Each ICC national/regional office is funded differently. The effectiveness of the international chair depends in large part on the state of their national office and their initial successes at fund-raising.

20. See in particular *Northern Lights against POPs* (Downie and Fenge 2003) for a full scientific and social explanation. Breast milk samples from Inuit women revealed some of the highest evidence of POPs in human life forms anywhere. Further research verified that POPs were traveling from specific factories in the United States, cementing global connectedness and the image of pollution circulating and being deposited in what was long thought of as a pristine world of snow and ice, far from the ills of industrial pollution in urban centers. In the *New York Times* story on this (Hilts 2000), Watt-Cloutier was quoted as "enthusiastically" saying that the study led by Dr. Barry Commoner put "names and faces to those who produce the dioxin that ends up in the north . . . so we can even call them up, visit them, and talk about what we are worried about." The article notes that 44,000 incinerators and factories are listed as sources for the dioxins.

21. Drawing on Italian philosopher Antonio Gramsci, cultural studies theorist Stuart Hall (in Maaka and Andersen 2006) introduced the notion of articulation in order to set aside questions of authenticity, particularly in relation to diasporic communities who argued both for continuity—the continual presence of historical, cultural, and economic relations with all of the attendant ruptures, inequities, and postcolonial symptoms, as well as new forms of political expression. Hall explains articulation as being like an articulated lorry—a truck with pieces added or subtracted from it. Such multi-piece articulation is evident in many of the quotes and positions articulated in this chapter—with sedimentation of the past and hopes for the future, as well as a pointed critique of the present bundled together. With regard to transnational indigenous movements in the Pacific, James Clifford generates a question that has distinct meaning in the Arctic as well: "In articulation theory, the whole question of authenticity is secondary, and the process of social and cultural persistence is political all the way back. It is assumed that cultural forms will always be made, unmade, and remade. Communities can and must reconfigure themselves, drawing selectively on remembered pasts. The relevant question is whether, and how, they convince and coerce insiders and outsiders, often in power charged, unequal situations, to accept the autonomy of a 'we.' . . . How should differently positioned authorities (academic and nonacademic, Native and non-Native) represent a living tradition's combined and uneven processes of continuity, rupture, transformation, and revival?" (Clifford 2001, 480). Self-determination, Clifford reminds us, is a multifaceted set of goals and representations that are evolutionary in nature.

Chapter Two:
Reporting on Climate Change

1. Schneider passed away July 20, 2010, and was fondly eulogized by many journalists and scientists. The *New York Times* carried his obituary.

2. See, e.g., Blum, Knudson, and Henig's 2006 *Field Guide for Science Writers*.

3. As articulated in the introduction, I am using this term in the book to explain the ways in which scientists and science journalists make sense of their obligations to professional norms and practices around objectivity, bias, and independence, and yet also see the need to intervene in public discourse in varied ways such that a rationale for publics (and their polities) to act emerges.

4. See also his essay with Jules Boykoff 2004; Brossard et al. 2004; Lahsen 2005b; 2010; Leiserowitz 2004; Mazur and Lee 1993; Oreskes and Conway 2010; Krosnick et al. 2006; Malka et al. 2009; Nisbet 2009; Nisbet and Scheufele 2009; Nisbet and Myers 2007; Scruggs and Benegal 2012; Ungar 1992.

5. See also Wynne 2008 for a corresponding call in STS.

6. The guide's three editions were funded by a combination of Department of Energy's Office of Science and NOAA.

7. When I spoke with Rensberger, he was nearing the end of a decade as the head of the Knight Science Journalism Fellowship Program at MIT. The *Washington Post* consolidated all of its science, environment, and health reporting under one editor in 2009, creating a new section that enrolls all three broad topics. It also signed a content-sharing agreement in 2008 with major online environmental source, Grist .org. (Russell 2009b).

8. The workshops were managed by the Metcalf Institute for Marine and Environmental Reporting and funded by EPA and NSF with in kind support from NOAA and NASA. They were held at universities in Rhode Island, San Diego, Seattle, New York, Berkeley, and Washington between 2003 and 2007.

9. I heard many of these concerns at the Society for Environmental Journalists' annual meeting at Stanford in the fall of 2007. There was a panel titled rather literally: "Journalists and Scientists: Can This Relationship Be Saved?" The panel included scientists who were part of the Aldo Leopold Program at Stanford as well as leading science journalists. The Leopold Program is a program for early or midcareer scientists to teach them how to talk to and think about media.

10. The same would apply to the more recent SuperStorm Sandy in 2012 or Typhoon Haiyan in 2013, which saw many of the same experts consulted and similarly quoted.

11. The "geopolitical" high stakes "game" that stands out the most was when the Russians sent a submarine down to plant their flag on an undersea continental shelf, claiming their territory (and the oil and gas therein).

12. Hoggan and Littlemore (2009) present investigative evidence and arguments that show how skeptics have intentionally targeted local and regional media outlets due to their lack of resources for science analysis and lack of savvy about discourse and tactics occurring in policy, science, and media arenas. I reviewed their book for *Nature* in early 2010.

13. Since I spoke with Dean, she has also published *Am I Making Myself Clear? A Scientist's Guide to Talking to the Public* (2009).

14. For most journalists, the practice of news is about "making sense" of events or findings and doing so in a way that compels readers to keep reading past the first line (the lead or lede) and the first paragraph (the nut graf). News stories follow a rather mechanistic formula called "an inverted pyramid" where the most important elements (reflecting core news values like impact, timeliness, proximity, prominence, conflict) go first and the details and context follow in descending order of importance. Peter Cole, a UK journalism professor writing for *The Guardian's* series on journalism, describes the process of writing news this way: "Journalists write stories for their readers to tell them what's going on, to inform them, engage them, entertain them, shock them, amuse them, disturb them, uplift them" (2008). Cole describes the work of journalism as "telling people what they didn't know and making them want to know it." Most readers/viewers/listeners have a well-developed sense of what news should feel/sound/look like, and when journalists break convention or apply their craft poorly, their audiences know it. These forms and styles are now being questioned by scholars and practitioners as new platforms emerge, but the inverted pyramid as Boczkowski and Mitchelstein (2013) show in *The News Gap* remains a stable and popular form even if audiences aren't paying as much attention to politics or public policy issues as they do to sports, crime, weather, and entertainment.

15. In my interview with him, Rensberger pointed out two major developments in science journalism—the splitting off of an environment beat and "the rise of advocacy groups as much more potent players in the public education scene." Advocacy groups are a major element to navigate for journalists reporting on climate change—and they've only begun to speak more loudly and forcefully with the rise of blogs and social media. Not only do advocates often lobby reporters and inundate newsrooms with press releases. Many also respond to stories either positively or negatively through blogs and other social media outlets like Twitter and Facebook. Rensberger pointed out that advocacy groups often present a "selected subset" of evidence where it supports their position on an issue, and Dean responded that she considers them a "news source," meaning those she reports on. But increasingly, as the robust blogosphere shows, advocates are part of a larger conversation with journalists and diverse publics.

16. See Rosen 1999, and Jenkins 2004 and 2012 with Sam Ford and Joshua Green.

17. Peters, in his analysis of *The Daily Show* in *Rethinking Journalism* (2013), argues that "it represents a critical, cultural pedagogy about the fundamental ethics of journalism." By focusing its critique on cable shows, *The Daily Show* identifies what journalism should be, making journalism accountable to its own norms.

18. See Brossard and Scheufele's op-ed "The Nasty Effect" in *The New York Times* on March 2, 2013, which *Popular Science* cites and quotes. The journal article it is based on is Anderson et al. (2013), "The Nasty Effect: Online Incivility and Risk Perceptions of Emerging Technologies," *Journal of Computer-Mediated Communication*.

19. Earlier in 2007, he had brought in Dr. Michael Crichton, author of the novel *State of Fear*, for testimony before the Senate committee that Inhofe chairs.

20. In my interview with Bud Ward, he indicated that these major environmental incidents were among the reasons why the Society for Environmental Journalists was originally formed.

21. In his 2009 book, *Global Journalism Ethics*, Ward argues that journalists should more fully embrace their roles as advocates for democracy. He also calls for a new and collaborative code of ethics for journalists to craft in concert with the public. Ethics are, Ward points out, a rhetorical framework that evolves over time, and new media have ushered in a significant moment of evolution. This doesn't mean ignoring the need for facts or declaration of opinion. Instead, Ward proposes pragmatic objectivity, a reflexive version of the professional norm of objectivity, as a way to address the expectation of impartiality and the responsibility inherent in the role of journalism.

22. See Callison 2012; 2009.

23. See for example the robust science communications discussions on Twitter at #scicomm, #scionline, #sciresearch, or conference specific hashtags associated with conferences like #scio14 or #AAASmtg. Other "classic" hashtags like #climate, #climatechange, and #COP19 don't always reflect these much more specific discussions about how science should be reported on and communicated to wide publics.

24. Shackley and Wynne found something similar among modelers in the mid-1990s, but this is not an oft-repeated criticism, as Revkin's story and the backlash to it demonstrate, and it doesn't mean that climate change doesn't pose enormous risks.

25. I will deal centrally with how scientists are navigating this in chapter 4, but understanding journalistic responses requires some reflection on how and what scientists are telling them about how to cover climate change. So while there's some overlap, I hope to minimize repetition by focusing in this chapter on the task journalists confront. I will return to the framework of "risk management" offered by Schneider both in chapters 4 and 5 (where I discuss Ceres's use of "climate risk").

Chapter Three:
Blessing the Facts

1. See DeWitt 2007a; Berry 2000; Bouma-Prediger 2001; Robinson 2007; Robinson and Chatraw 2006; Sleeth 2007.

2. During the summer of 2008, as I was beginning to analyze my fieldwork, I often took the subway between the Harvard Square and Kendall-MIT stops. A local church had placed the last line on a prominent billboard at the Harvard "T" stop, and it remains there as of 2013. Upon looking up the quote, I was surprised to find it part of a much larger discussion about how and where science fits in the Christian worldview.

3. Joel Hunter has been referred to as the spiritual advisor to President Obama. He gave the prayer at Obama's victory party following his election. For one example of news coverage, see "Where the President Turns for Spiritual Advice: Rev. Joel Hunter of Longwood, FL" (14 April 2009) from *Black Christian News* at http://www.blackchristiannews.com/news/2009/04/where-the-president-turns-for-spiritual-advice-rev-joel-hunter-of-longwood-fl.html.

4. Katharine Wilkinson's recently published *Between God and Green* (2012) provides a thorough and deeply historicized analysis of "evangelical climate care" in light of these larger changes, drawing on focus groups, interviews, and some of the events I recount here, including the launch of the Evangelical Climate Initiative. Our goals are somewhat similar in elucidating how it is that climate change comes to have meaning for American evangelicals, but this chapter focuses more directly on how epistemological differences are addressed and when and how they matter in terms of investing science with meaning, ethics, and morality.

5. The Vineyard church movement began in the 1970s in Yorba Linda, California, and is something like a denomination, but is sometimes referred to as "nondenominational" for its independence from other evangelical groups. According to the Vineyard USA website, there are over 1,500 Vineyard churches worldwide and 550 in the United States.

6. See DeWitt 2007b for an in-depth account of the groundwork laid prior to the 1990s.

7. Cizik said this description of a "conversion" was controversial among con-

servatives when he described it in retrospect during a 2010 panel on "The Cost of Conscience: Dissent in the Workplace—A Conversation with Matthew Alexander, Richard Cizik, Elizabeth MacKenzie Biedell, and Morton H. Halperin," sponsored by Open Society Foundations and broadcast on Fora.tv; http://fora.tv/2010/05/11/The_Cost_of_Conscience_Dissent_in_the_Workplace.

8. *Time* reported that same year that "season creep" was a common "neologism" explaining the cause of early spring with this line: "Most scientists say it's global warming" (Sayre 2006). Most scientists I encountered would be hard pressed to make such a statement, since attributing anything directly to climate change, and in particular with regard to weather anomalies, is quite difficult. Longer term trending can be seen as related to climate change, however.

9. Open Society Fellowships are funded by George Soros, a well-known liberal and Democrat.

10. Cizik was recently profiled along with Hunter in *America's New Evangelicals* by Marcia Pally as evangelical leaders who have "left the Right."

11. This analysis is reflected in the topics and themes that *Creation Care*, the magazine, deals with as well. *Creation Care* is put out by EEN. More recently, EEN has been working with environmental justice advocates in Appalachia, which perhaps disrupts notions of the middle-class white evangelical as primary focus.

12. Additionally, new submovements like the emerging church and other splinter groups are reshaping the conversation, rallying like minds through blogs and books beyond even what mega-churches are capable of doing—Brian McLaren's "Everything Must Change" book and its tour across the United States throughout 2008 and 2009 being one prominent example where the environment was introduced as a new and pressing priority for Christians. Cizik and Ball were both clear that Creation Care is not attached or associated with the emergent church.

13. When *Science* reported on ECI, they quoted Jim Furnish, former deputy chief of the U.S. Forest Service, as saying, "What's going on here is peacemaking at its most basic level between the religious and scientific worldview" (Kintisch 2006). The article stated that the ECI was the culmination of "a 5-year effort by a handful of scientists, most of them devout Christians, to find common ground with an influential Republican constituency that is often an implacable enemy in science policy debates." The article featured a photo of Cizik, DeWitt, and Houghton with the caption "Warming Trend."

14. This paragraph footnotes: ["For the Health of the Nation: An Evangelical Call to Civic Responsibility," approved by National Association of Evangelicals, October 8, 2004.]

15. Other declarations have been launched by the Cornwall Alliance since the period surrounding ECI, and the organization remains active while ECI's website is no longer operational and EEN no longer mentions ECI on its site.

16. Not only that, but Gore's affiliation with the Democratic Party has made him a less persuasive spokesperson for climate change among evangelicals as well. Cizik explained to me that they tried to give tickets away to *An Inconvenient Truth*,

but evangelicals wouldn't take them. So they sent out and recommended that evangelicals watch *The Great Warming*, also released in 2006, narrated by Keanu Reeves and Alanis Morissette, and produced by a Canadian company. It reiterates much of the scientific evidence, but also features interview clips with Cizik and other faith leaders, as well as activists and "real" people.

17. Here are the lyrics to the first verse:

I come to the garden alone,
While the dew is still on the roses,
And the voice I hear, falling on my ear,
The Son of God discloses.
And He walks with me, and He talks with me,
And He tells me I am His own,
And the joy we share as we tarry there,
None other has ever known.

18. Ball also went on to note that this problem of trust in science is widespread beyond the bounds of faith where it seems that scientists are constantly issuing new research findings that may or may not contradict previous findings—the "everything good is now bad for you" problem that is especially prevalent in medicine. He used the example of cholesterol. Yet that is qualitatively different than the problem expressed regarding morality and sexual orientation.

19. At the Creation Care conference, there were several information and display tables set up around the lunch area. Most of them were for organizations like A Rocha: Christians in Conservation or the Au Sable Institute as well as Christian publishers with many new titles on environmental themes. A couple of the tables were for secular environmental groups. At the Sierra table, I met Lyndsay Moseley, a Christian and Sierra Club employee. She reiterated to me what Richard Cizik had said in my interview with him—that evangelicals had to be extremely careful about forming partnerships with secular groups at this stage in the development of Creation Care. It would be too easy to write off their efforts if they were perceived as liberal, leftist, or secular. After the conference, Moseley edited a book that Sierra published: *Holy Ground: A Gathering of Voices on Caring for Creation* (2008).

Chapter Four:
Negotiating Risk, Expertise, and Near-Advocacy

1. The most reported debate was among economists regarding the discount rate, an economic term denoting how models comparatively value present and future costs and benefits. Citing ethical grounds, Stern used a near-zero rate to compare dollars spent now on emissions reductions with dollars in the future. Yale economics professor William Nordhaus disagreed vehemently and very publicly with Stern, and their debate at Yale in February 2007 was reported on by the *New York Times* as a "juicy" academic fight with public policy ramifications. Nordhaus argued that a 3 percent discount rate is more palatable—that "benefits accrued in twenty-five years' time are worth about half their current value." The idea behind this valuation is that it's better to do less now because in the future we'll all be richer and able to cope better with whatever warming brings.

2. The full list as advertised at that time: Moderated by Professor Robert M. Solow, Institute Professor and Professor of Economics Emeritus. Panelists include Professor Paul L. Joskow, MIT Department of Economics; Professor Stephen Ansolabehere, MIT Department of Political Science; Dr. A. Denny Ellerman, MIT Center for Energy and Environmental Policy Research; Professor Henry Jacoby, Sloan School of Management and MIT Joint Program on the Science and Policy of Global Change; Professor Ronald Prinn, Department of Earth, Atmospheric, and Planetary Sciences and Center for Global Change Science; Dr. John Reilly, MIT Joint Program on the Science and Policy of Global Change; and Dr. John Parsons, MIT Center for Energy and Environmental Policy Research.

3. Most scientists work with a sliding scale of certainty. This became evident in the IPCC's fourth assessment report, in which they began talking about likely and very likely scenarios. Similarly, the ACIA used such language. It doesn't include "outliers," changes that are faster or slower than the expected range of rates. During fieldwork, I heard two talks about the possibility of "abrupt climate change" from paleoclimatologists based on their reading of evidence from the distant past, but they are not generally mentioned as within the usual range of potential scenarios. I quoted Stephen Schneider in chapter 2 saying that scientists "worry endlessly about the tails" on probabilities, and this is another example of that problematic.

4. In a 2007 lecture at MIT, speaking as he was to peers in the Joint Change program run by Prinn and Jacoby, Schneider praised them for their interdisciplinarity— meaning science (atmospheric, ocean, earth) and economic modeling along the lines of the *Stern Review*. Schneider said this was the future for climate change where scientific findings and economics must work together, but that it was a difficult partnership to do well. He cited MIT's program as one of the few models he could point to where it was being done well.

5. Here, we might also think of the effects of "right to know" legislation on toxicology, for example (Fortun and Fortun 2005).

6. Joseph Dumit (2012) has shown the way this logic has been applied by the pharmaceutical industry to create a "common sense" regimen of health practices that correspond to statistics rather than evidence. Risk and insurance then are malleable terms and forms of life that have been repurposed in more than one area of scientific discourse, and what these terms mean in any given situation can be understood through how they are used and what grammars and rules grow up around their use.

7. NYU professor and journalist Charles Seife (2012), in his popular book *Proofiness*, says: "We are prey to risk management" because "minor changes in wording can easily make a huge risk seem worth taking or an insignificant risk seem dangerous. As a result we are vulnerable to manipulation. We can't easily detect

when someone is understating or exaggerating risks." Seife cites episodes in which government agencies, media, and financial institutions have led the public astray by overstating and understating the risks associated with decisions. This kind of out-of-reach discussion, which is only possible among experts familiar with the objective instruments of choice, that is, numbers, is what early American philosophers John Dewey and Walter Lippmann debated as well when they saw the role of experts become central to media and democracy.

8. Advocacy is a transgression of professional norms for scientists and constitutes what Ludwig Fleck (1979) has called a "slogan word" that has acquired a "magical power" either for or against a word's intended meanings and applications. Near-advocacy, in contrast to advocacy, implies this other action of turning findings into numbers, of finding ways to make findings and predictions into interventions that might have some traction in iterations of what climate change is and what should be done to avert its inherent risks.

9. Roger Pielke (2007), a climate policy analyst and political scientist, has suggested that scientists have generally acted in four roles: pure scientist, science arbiter, issue advocate, and honest broker. He suggests that honest brokers use their expertise to help policymakers understand the full range of policy alternatives. In so doing, he reserves a particular place for expertise within the political process and a decidedly nonactivist role for scientists. Mike Hulme (2009), a climate science expert, while sympathetic to Pielke's suggestion, notes the impossibility of remaining only an "honest broker" and being perceived as such.

10. First theorized in 1956 by Marion King Hubbert, "*Peak oil* is a term that summarizes the concept that the production of crude oil—as well as that of most finite resources in a market economy—grows, reaches a maximum (peak), and then gradually declines to zero" (Bardi 2009, 323). Hubbert's theory was applied to the U.S. with the peak occurring in 1971, and has been applied by analysts to world oil production with predictions of its peak sometime in the early twenty-first century. See Bardi 2009 for an overview of the theory, its applications, and responses to it.

11. Science and Democracy Lecture and Panel Discussion: Professor Yaron Ezrahi, Hebrew University of Jerusalem, "Necessary Fictions: Imagining Democracy after Modernity," April 9, 2007, Starr Auditorium, Kennedy School of Government. Panelists: Ellen Goodman, *Boston Globe*, and Fellow, Shorenstein Center, KSG; James McCarthy, Organismic and Evolutionary Biology, Harvard University; Steven Shapin, History of Science, Harvard University; Cass Sunstein, Chicago and Harvard Law Schools.

12. Yet, Besley and Nisbet (2013) have argued based on data collected at AAAS meetings that the majority of scientists are not averse to speaking with media, and see it as a necessary avenue for reaching publics. Based on the same data set, Besley, Oh, and Nisbet (2013) argue that this is more likely to be a mid-career step for scientists and varies according to field of study (chemists are the least likely to speak with media).

13. Dean doesn't call climate "skeptics" by that term because she feels so strongly

about science as an enterprise of skepticism. Instead, she refers to skeptics as "dissidents." I've stuck with *skeptic* here because it is the most commonly used term, but Dean raises an excellent point about language.

14. Mooney, too, pointed out that science reporters should not be included in such characterizations because they often "love" science, are careful about representing findings, and have enormous respect for their sources.

15. In their ethnography of contemporary U.S. toxicology, Fortun and Fortun (2005) explain emergent forms of life in the sciences as related explicitly to "the experimental traditions and systems in which scientists conduct their work" (43). Drawing on work by Hans Jorg Rheinberger, who has described the technical conditions at work in experimental systems, Fortun and Fortun introduce the concept of "civic science." It is in some ways nested in the concept of civic epistemologies, but it also works to explain more specifically the work that scientists understand themselves to be doing both in advisory roles and more generally in terms of how they perceive their contributions and obligations to the "public good." Civic science accounts for what "good science" is and how it is enunciated, situating it as "a historical effect, produced by a tangle of social, political, technological, and biomaterial forces." And methodologically in ethnographic terms, it also seeks to understand what scientists see as "worthy of care and ethical attention"—a facet of particular importance given the new forms of data and informational infrastructure that have transformed all aspects of toxicology.

Much like articulations of "good journalism," Fortun and Fortun note that what is enunciated as "good science" is "rooted in tradition of thought and practice, even when intended to establish new agendas and open up new lines of work" (2005, 47). This, wherever it is elaborated, creates tension between "testing and experimentation" or between "verification of knowledge and the production of fundamentally new knowledge." And it's in this crux—what they call a double bind, much like the one that underlies this book—that ethics resides. Within the constraints and terrain of tradition, ethical decisions are enacted at every stage of the scientific process in order for experimental systems to both forge ahead and function as a "research tool" and maintain fidelity to methodologies. This tension is what produces the enunciations of "good science" and what Michael M. J. Fischer (2003) calls "ethical plateaus," "where multiple technologies interact to create complex terrain or topology of perception and decision making" and the sedimentation of past decisions and persistent double binds accrete.

Chapter Five:
What Gets Measured Gets Managed

1. The roster includes some of the largest and most influential publicly held corporations in the United States. A sample includes PepsiCo, Bank of America, Sodexo, Virgin America, Time Warner, Sunoco, PG&E, Gap, Exelon, GM, General Mills, Levi Strauss, McDonald's, and Nike. Smaller influential companies are also listed, including Native Energy, Interface, Seventh Generation, and Green Mountain Coffee Roasters.

2. This research was undertaken at the beginning of an economic recession that defiantly persisted for years and prior to major changes in banking and the BP "Deep Horizon" oil spill. As noted in the introduction, this research is meant to serve as a window into the evolution of climate change's form of life as it is articulated and substantiated in these groups.

3. Ceres founder Joan Bavaria said the same thing in her obituary video on the Ceres site.

4. She later told an MIT Sloan dissertation student that if companies had been involved, they probably never would have reached an agreement. It was hard enough to come to an agreement with the nonprofits and SRIs in the room.

5. ICCR is not related to Creation Care, although they are aware of one another and seem to share some joint involvement with third-party ecumenical efforts such as the National Religious Partnership for the Environment (NRPE). ICCR is an agglomeration of mainline Protestant and Catholic organizations and known for its shareholder activism on a range of issues relating to social and environmental concerns.

6. The American Federation of Labor and Congress of Industrial Organizations (AFL-CIO), the largest federated union in the United States, is also a key player in shareholder activism on labor issues.

7. Bavaria passed away following a long battle with cancer later in 2008.

8. See Fifka 2011; Cormier, Ledoux and Magnan 2011; Vurro and Perrini 2011 for recent overviews of the contributions of social and environmental reporting for corporations.

9. Ceres's membership list of over fifty environmental and public interest organizations includes AFL-CIO, Rainforest Alliance, Sierra Club, Earthwatch Institute, Oxfam, and Union of Concerned Scientists. Their longer list of foundations and investors includes the Evangelical Lutheran Church in America, the New York State Comptroller's Office, Trillium Asset Management Corporation, Calvert Group, and the California State Treasurer's Office. It is an agglomeration of these members as deemed relevant to the corporations that compose its stakeholder group.

10. In 2009, a large group of environmental groups around the world came together to call for an end to the oil sands development in northern Alberta, Canada. Even a former Alberta premier, Peter Lougheed, who presided over an earlier oil boom in the province, has called for a moratorium on the development, citing enormous environmental damage already incurred by the project in its early phases. See Nikiforuk 2010.

11. Bob Massie, in his talk at Sloan, pointed out that with GRI, many were afraid it was "a plan for the world done by Ceres," so Ceres decided to spin it off. He pointed out that this is common in the corporate world, but not so common among nonprofits.

12. See, e.g., the Carbon Disclosure Project, which involves companies like Dell. Ceres points out in its literature that it's a good option but a very narrow expression of sustainability reporting, focusing only on carbon emissions.

13. GRI in its G3 iteration divides disclosure into three types: (1) profile that covers strategy and analysis, (2) management approach, and (3) performance indicators. There are nine economic indicators (which include financial/material impacts of climate change on business), thirty environmental indicators (fifteen of which are core like water usage), fourteen labor (Snyder noted these do not have the same maturity as the environmental categories), nine human rights ("these enjoy the least amount of consensus that we've got them right. It's the best we can do right now," Snyder said), eight society indicators (six are core), and seven product responsibility (four are core). Depending on application levels, Snyder told an interested group at the 2007 conference, companies receive a grade and extra merit if it has been externally reviewed. So a C–grade would be for ten indicators, B for twenty indicators, and A for fifty indicators including core, sectors, and management approach. Part of the GRI process also includes mapping stakeholders, and Snyder called it an "iterative process." In the question period, Snyder agreed that while a financial report means reporting on "what you own," a sustainability report must factor in additional issues, particularly in the case of multinationals. What a company controls, owns, or influences is difficult to demarcate in joint ventures and globalized locations.

14. For more, see Catherine Brahic, "What a Slump in Carbon Prices Means for the Future," *New Scientist*, 11 February 2009.

15. See, e.g., the Durban Group for Climate Justice at www.durbanclimatejustice .org/.

16. GE's Ecoimagination initiative was about transforming their investments in research and development of alternative energy technology to the tune of $1.5 billion by 2010 (up from $400 million in 2005), but it also has a policy and public engagement aspect to it.

17. See Oreskes and Conway 2010 and Hoggan and Littlemore 2010 for more on this.

18. Gore also spoke at the 2006 Ceres conference and presented his slide show there.

19. The crisis managed to force into bankruptcy or force low-priced acquisition ("fire sale") of three major investment banks: Lehman Brothers, Merrill Lynch, and Bear Stearns. The two remaining major investment banks, Goldman Sachs and Morgan Stanley, agreed to become commercial banks and face more regulation. The credit crisis didn't just engulf the financial sector. Iceland was forced to declare bankruptcy, and many other smaller countries were also hit extremely hard with investments tied up in one or another to what was happening with American mortgages.

20. Swiss Re was a major sponsor of *The Great Warming*, the film used by evangelicals as an alternative to Al Gore's *Inconvenient Truth*. Ceres's report "From Risk to Opportunity: How Insurers Can Proactively and Profitably Manage Climate Change" names other actions taken by Munich Re, Lloyd's of London, Allianz, and others.

21. When I checked in with Kelly in 2008, she said that Marsh had changed CEOs, and it was no longer sponsoring the Forum, so they were looking for an alternative partner at that time, but in searching their website recently, Marsh is still listed as a sponsor.

22. USCAP stands for United States Climate Action Partnership, which describes itself on its website as "a group of businesses and leading environmental organizations that have come together to call on the federal government to quickly enact strong national legislation to require significant reductions of greenhouse gas emissions." Its website does not have any activity past 2011.

Epilogue:
Rethinking Public Engagement and Collaboration

1. In anthropological terms, a para-site was constructed by way of a student-organized conference at MIT in April 2008, titled "Disruptive Environments: Activists and Academics in Conversation." The panel I organized on climate change featured many of my key informants in conversation with one another: Boyce Rensberger, Kerry Emanuel, Andrew Revkin, Naomi Oreskes, and Kevin Conrad. George Marcus (2000) uses the term *para-site* to refer to spaces for interaction, collaboration, and reflection that are consciously constructed and orchestrated by both researchers and subjects/informants. Para-sites reflect "the reality of fieldwork as movement in complex, unpredictable spatial and temporal frames" and create "space outside conventional notions of the field in fieldwork to enact and further certain relations of research essential to the intellectual or conceptual work that goes on inside such projects." These spaces have often existed informally in anthropological fieldwork where subjects, patrons, and researchers come together in order that dialogue expression might allow for ideas to be tested and mistakes corrected.

References

Achirgina-Arsiak, T. 1992. "Paianitok!" *Étude/Inuit/Studies* 16: 47–50.

Alia, V. 1999. *Un/Covering the North: News, Media, and Aboriginal People.* Vancouver: UBC Press.

Alia, V. 2007. *Names and Nunavut: Culture and Identity in the Inuit Homeland.* New York: Berghahn.

Anders, G., and S. J. Langdon. 1989. "Social and Economic Consequences of Federal Indian Policy: A Case Study of the Alaska Natives." *Economic Development and Cultural Change* 37:285–303.

Anderson, A. 2008. "Short Sharp Science: Get Ready for the Inuit Oil Millionaires." *New Scientist,* September 24. http://www.newscientist.com/blogs/shortsharpscience/2008/09/why-inuits-will-make-great-oil.html.

Anderson, A. A., D. Brossard, D. A. Scheufele, M. A. Xenos, and P. Ladwig. 2013. "The 'Nasty Effect': Online Incivility and Risk Perceptions of Emerging Technologies." *Journal of Computer-Mediated Communication* 19, no. 3: 373–87.

Anderson, B. 1991. *Imagined Communities: Reflections on the Origin and Spread of Nationalism.* London: Verso.

Arnold, R. 1978. *Alaska Native Land Claims.* Anchorage: Alaska Native Foundation.

Bailey, R. 2009. "Commentary: John Holdren, Ideological Environmentalist." *Forbes,* February 3. http://www.forbes.com/2009/02/03/holdren-obama-science-opinions-contributors_0203_ronald_bailey.html.

Ball, J. 2007a. "In Climate Controversy, Industry Cedes Ground." *Wall Street Journal*, January 23. http://online.wsj.com/news/articles/SB116949687307684055.

Ball, J. 2007b. "Ungodly Distortions: Evangelical Christians Know That Caring for God's Creation Is a Scriptural Imperative." Beliefnet. http://www.beliefnet.com /News/2005/03/Ungodly-Distortions.aspx?p=1.

Ball, J. 2008. "Hope Springs Eternal." Deep Green Conversation. http://deepgreen conversation.org/index.php?s=cizik.

Banerjee, N. 2006. "Pastor Chosen to Lead Christian Coalition Steps Down in Dispute over Agenda." *New York Times*, November 28. http://www.nytimes.com/2006 /11/28/us/28pastor.html?_r=0.

Banerjee, N. 2008. "Taking Their Faith, but Not Their Politics, to the People." *New York Times*, June 1. http://www.nytimes.com/2008/06/01/us/01evangelical.html ?pagewanted=all.

Bardi, U. 2009. "Peak Oil: The Four Stages of a New Idea." *Energy* 34, no. 3: 323–26.

Beck, U. 1992. *Risk Society: Towards a New Modernity*. Thousand Oaks, CA: Sage.

Beck, U. 2002. "The Silence of Words and Political Dynamics in the World Risk Society." *Logos* 1, no. 4: 1–18.

Beisner, E. C. 2007. *NAE, or Cizik, Off Course: The National Association of Evangelicals Is Leading Its Member Churches into the Political Wilderness*. http://www.cornwall alliance.org/blog/item/nae-or-cizik-off-course/.

Benjamin, W. 1968. "The Task of the Translator." In *Illuminations*, ed. H. Arendt. New York: Schocken Books.

Benkler, Y. 2006. *The Wealth of Networks: How Social Production Transforms Markets and Freedom*. New Haven, CT: Yale University Press.

Benson, E. 2008. *The Wired Wilderness: Electronic Surveillance and Environmental Values in Conservation Biology*. Cambridge, MA: MIT Press.

Benson, R., and E. Neveu. 2005. *Bourdieu and the Journalistic Field*. Malden, MA: Polity Press.

Berger, E. 2008. "Hurricane Expert Reconsiders Global Warming's Impact." *Houston Chronicle*, April 11.

Berkes, F. 1977. "Fishery Resource Use in a Subarctic Indian Community." *Human Ecology* 5:289–307.

Berkes, F. 1993. "Traditional Ecological Knowledge in Perspective." *Traditional Ecological Knowledge: Concepts and Cases*, ed. J. T. Inglis, 1–9. Ottawa: Canadian Museum of Nature and the International Development Research Centre.

Berry, R., ed. 2000. *The Care of Creation: Focusing on Concern and Action*. Westmont, IL: InterVarsity.

Besley, J. C., and M. Nisbet. 2013. "How Scientists View the Public, the Media and the Political Process." *Public Understanding of Science* 22, no. 6: 644–59.

Besley, J. C., S. H. Oh, and M. Nisbet. 2013. "Predicting Scientists' Participation in Public Life." *Public Understanding of Science* 22, no. 8: 971–87.

Bloor, David. 1983. *Wittgenstein: A Social Theory of Knowledge*. New York: Columbia University Press.

Blum, D., M. Knudson, and R. M. Henig, eds. 2006. *A Field Guide for Science Writers:*

The Official Guide of the National Association of Science Writers. New York: Oxford University Press.

Blunt, S. H. 2006. "The New Climate Coalition: Evangelical Leaders Bolster the Fight against Global Warming." *Christianity Today*, February 8. http://www.christianity today.com/ct/2006/februaryweb-only/106–34.0.html.

Boczkowski, P. 2004. *Digitizing the News: Innovation in Online Newspapers*. Cambridge, MA: MIT Press.

Boczkowski, P., and E. Mitchelstein. 2013. *The News Gap*. Cambridge, MA: MIT Press.

Bodenhorn, B. 2003. "Fall Whaling in Barrow, Alaska: A Consideration of Strategic Decision-Making." In *Indigenous Ways to the Present: Native Whaling in the Western Arctic*, ed. A. P. McCartney. Edmonton: Canadian Circumpolar Press Institute.

Boler, M., ed. 2008. *Digital Media and Democracy: Tactics in Hard Times*. Cambridge, MA: MIT Press.

Boroditsky, L. 2001. "Does Language Shape Thought?: Mandarin and English Speakers' Conceptions of Time." *Cognitive Psychology* 43, no. 1: 1–22.

Bouma-Prediger, S. 2001. *For the Beauty of the Earth: A Christian Vision for Creation Care*. Grand Rapids, MI: Baker Academic.

Bourdieu, P. 1991. *Language and Symbolic Power*. Cambridge, MA: Harvard University Press.

Boykoff, M. T. 2011. *Who Speaks for the Climate? Making Sense of Media Reporting on Climate Change*. New York: Cambridge University Press.

Boykoff, M. T., and J. M. Boykoff. 2004. "Balance as Bias: Global Warming and U.S. Prestige Press." *Global Environmental Change* 14:125–36.

Brahic, C. 2009. "What a Slump in Carbon Prices Mean for the Future." *New Scientist*, February 11. http://www.newscientist.com/article/dn16583-what-a-slump -in-carbon-prices-means-for-the-future.html#.UznSM8f9ojE.

Broder, J. M. 2008. "Obama Courting Evangelicals Once Loyal to Bush." *New York Times*, July 1. http://www.nytimes.com/2008/07/01/us/politics/01evangelicals .html?pagewanted=all.

Brody, H. 1991. *The People's Land: Inuit, Whites, and the Eastern Arctic*. Vancouver: Douglas and McIntyre.

Brody, H. 1997. *Maps and Dreams*. Long Grove, IL: Waveland Press.

Broersma, M. 2010. "The Unbearable Limitations of Journalism: On Press Critique and Journalism's Claim to Truth." *International Communication Gazette* 72, no. 1: 21–33.

Brossard, D., J. Shanahan, and K. McComas. 2004. "Are Issue-cycles Culturally Constructed? A Comparison of French and American Coverage of Global Climate Change." *Mass Communication and Society* 7, no. 3: 359–77.

Brossard, D., and D. A. Scheufele. 2013. "This Story Stinks." *New York Times*, March 2. http://www.nytimes.com/2013/03/03/opinion/sunday/this-story-stinks.html.

Burri, R. V., and J. Dumit. 2007. *Biomedicine as Culture: Instrumental Practices, Technoscientific Knowledge, and New Modes of Life*. New York: Routledge.

Buse, U. 2007. "Is the IPCC Doing Harm to Science?" *Der Spiegel*, May 3. http://

www.spiegel.de/international/world/emotionalizing-climate-change-is-the-ipcc
-doing-harm-to-science-a-480766.html.

Bush, V. 1945. "Science: The Endless Frontier: A Report to the President by Vannevar Bush, Director of the Office of Scientific Research and Development, July 1945." Washington, DC: U.S. Government Printing Office.

Callison, C. 2009. "Engagement and the New Media: Social Affiliations Provide More than a Passing Interest in How We Act on Issues." *Vancouver Sun*, February 12. http://www2.canada.com/vancouversun/news/westcoastnews/story
.html?id=7800afoa-04ab-4eaf-ba8f-5a8abe17c7f7&p=2.

Callison, C. 2012. "New Media Means New Ways to Cover Climate Change." *J-source.ca*, February 15. http://j-source.ca/article/new-media-means-new-ways-cover-climate
-change.

Campolo, T. 2008. *Red Letter Christians*. Ventura, CA: Regal.

Carlson, J. 2006. "What Would Jesus Do about Global Warming?" *Orlando Sentinel*, April 13. http://www2.orlandoweekly.com/features/story.asp?id=10622.

Carroll, J. 2008. "Tillerson to Face Rockefellers, Nuns in Exxon Meeting Showdown." *Bloomberg*, May 27. http://www.bloomberg.com/apps/news?pid=newsarchive
&sid=aUHNmQ9ZCUzo.

"Carved Hares and Dancing Bears: Our Correspondent Explores the Arctic Art-World." 2007. *Economist*, June 22.

Castells, M. 2000. *The Rise of the Network Society*. Malden, MA: Blackwell.

Castells, M. 2003. *The Power of Identity*. Malden, MA: Blackwell.

Castells, M. 2005. *The Network Society: A Cross-Cultural Perspective*. London: Edward Elgar.

Castells, M. 2009. *Communication Power*. Oxford: Oxford University Press.

Ceres. 2006. "Managing the Risks and Opportunities of Climate Change: A Practical Toolkit for Corporate Leaders." http://www.ceres.org/resources/reports
/climate-risk-toolkit-for-corporate-leaders-2006/view.

Ceres. 2008. "Ceres Honors the Life of Founder Joan Bavaria (1943–2008)." http://
www.ceres.org.

Ceres. 2010. "Ceres Coalition." http://www.ceres.org/Page.aspx?pid=425.

Choy, T. 2011. *Ecologies of Comparison: An Ethnography of Endangerment in Hong Kong*. Durham, NC: Duke University Press.

Clifford, J. 2001. "Indigenous Articulations." *Contemporary Pacific* 13:467–90.

Collins, H. M., and R. Evans. 2002. "The Third Wave of Science Studies: Studies of Expertise and Experience." *Social Studies of Science* 32:235–96.

Cole, P. 2008. "How Journalists Write." *Guardian*, September 25. http://www.the
guardian.com/books/2008/sep/25/writing.journalism.

Colson, C. 2009. "Global Warming as Religion: You Had to See It Coming." *Breakpoint: Changing Lives, Minds, and Communities through Jesus Christ*. Accessed August 20, 2010. http://www.breakpoint.org/commentaries/13681-global-warming
-as-religion.

Cook, T. E. 1998. *Governing with the News: The News Media as a Political Institution*. Chicago: University of Chicago Press.

Couzin, J. 2007. "Polar Science: Opening Doors to Native Knowledge." *Science* 315 (March 16): 1518–19.

Cox, D. 2007. "Young White Evangelicals: Less Republican, Still Conservative." *Pew Forum Religion and Public Life Project*, September 28. http://www.pewforum.org/2007/09/28/young-white-evangelicals-less-republican-still-conservative/.

Crouch, A. 2006. "Letter to a Tenured Professor." *Christianity Today*. http://www.christianitytoday.com/ct/2006/septemberweb-only/139=22.0.html.

Cruikshank, J. 1991. *Reading Voices/Dan Dha Tsedeninth'e: Oral and Written Interpretations of the Yukon's Past*. Vancouver: Douglas and MacIntyre.

Cruikshank, J. 2001. "Glaciers and Climate Change: Perspectives from Oral Tradition." *Arctic* 54:377–93.

Cruikshank, J. 2005. *Do Glaciers Listen? Local Knowledge, Colonial Encounters, and Social Imagination*. Vancouver and Seattle: UBC Press and University of Washington Press.

Cormier, D., M. J. Ledoux, and M. Magnan. 2011. "The Informational Contribution of Social and Environmental Disclosures for Investors." *Management Decision* 49, no. 8: 1276–1304.

D'Angelo, Paul. 2002. "News Framing as a Multi-Paradigmatic Research Program: A Response to Entman." *Journal of Communication* 52:870–88.

Dahl, J., J. Hicks, and P. Jull. 2000. *Nunavut: Inuit Regain Control of Their Lands and Their Lives*. Copenhagen: International Work Group for Indigenous Affairs (IWGIA).

Daley, P. J., and B. A. James. 2004. *Cultural Politics and the Mass Media: Alaska Native Voices*. Champaign: University of Illinois Press.

Daly, R. 2005. *Our Box Was Full: An Ethnography for the Delgamuukw Plaintiffs*. Vancouver: UBC Press.

Damas, D. 1985. *Handbook of North American Indians*. Vol. 5, *Arctic*. Washington, DC: Smithsonian Institution Scholarly Press.

Dean, C. 2009. *Am I Making Myself Clear? A Scientist's Guide to Talking to the Public*. Cambridge, MA: Harvard University Press.

Deleuze, G. 1989. *Cinema 2: The Time-Image*. Minneapolis: University of Minnesota Press.

Deleuze, G., and F. Guattari. 1987. *A Thousand Plateaus: Capitalism and Schizophrenia*. Minneapolis: University of Minnesota Press.

Deutsch, C. H. 2008. "Saving the Planet? Not with My Money." *New York Times*, March 26.

Dewey, J. 1927. *The Public and Its Problems*. New York: Henry Holt.

DeWitt, C. 2007a. *Earth-Wise: A Biblical Response to Environmental Issues*. Grand Rapids, MI: Faith Alive Christian Resources.

DeWitt, C. 2007b. "Evangelical Environmentalism in America." In *Witnessing in the Midst of a Suffering Creation*, ed. L. Vischer, 174–204. Geneva: John Knox International Reformed Centre.

DiMento, J. F. C., and P. Doughman, eds. 2007. *Climate Change: What It Means for Us, Our Children, and Our Grandchildren*. Cambridge, MA: MIT Press.

Domingo, D., and A. Heinonen. 2008. "Weblogs and Journalism: A Typology to Explore the Blurring Boundaries." *Nordicom Review* 29:3–15.

Downie, D. L., and T. Fenge, eds. 2003. *Northern Lights against POPs: Combatting Toxic Threats in the Arctic*. Montreal: McGill-Queen's University Press.

Dreifus, C. 2006. "A Conversation with—Kerry Emanuel; With Findings on Storms, Centrist Recasts Warming Debate." *New York Times*, January 10.

Dumit, J. 2003. "Is It Me or My Brain? Depression and Neuroscientific Facts." *Journal of Medical Humanities* 24, nos. 1–2: 35–47.

Dumit, J. 2004a. *Picturing Personhood: Brain Scans and Biomedical Identity*. Princeton, NJ: Princeton University Press

Dumit, J. 2004b. "Writing the Implosion: Teaching the World One Thing at a Time." Unpublished article.

Dumit, J. 2012. *Drugs for Life: How Pharmaceutical Companies Define Our Health*. Durham, NC: Duke University Press.

Edwards, P. N. 1999. "Global Climate Science, Uncertainty and Politics: Data-Laden Models, Model-Filtered Data." *Science as Culture* 8:437–72.

Edwards, P. N. 2010. *A Vast Machine: Computer Models, Climate Data, and the Politics of Global Warming*. Cambridge, MA: MIT Press.

Eilperin, J., and J. Achenbach. 2008. "Advocates for Action on Global Warming Chosen as Obama's Top Science Advisers." *Washington Post*, December 19.

Emanuel, K. 2005. "Increasing Destructiveness of Tropical Cyclones over the Past 30 Years." *Nature* 436:686–88.

Emanuel, K., R. Sundararajan, and J. Williams. 2008. "Hurricanes and Global Warming: Results from Downscaling IPCC AR4 Simulations." *Bulletin of the American Meteorological Society* 89:347–67.

Entman, R. M. 1993. "Framing: Toward Clarification of a Fractured Paradigm." *Journal of Communication* 43:51–58.

Epstein, S. 1996. *Impure Science: AIDS, Activism, and the Politics of Knowledge*. Berkeley: University of California Press.

Eythorsson, E. 1993. "Sami Fjord Fishermen and the State: Traditional Knowledge and Resource Management in Northern Norway." *Traditional Ecological Knowledge: Concepts and Cases*, ed. J. Inglis. Ottawa: International Development Research Centre.

Ezrahi, Y. 1990. *The Descent of Icarus: Science and the Transformation of Contemporary Democracy*. Cambridge, MA: Harvard University Press.

Faubion, J., and G. Marcus. 2009. *Fieldwork Is Not What It Used to Be: Learning Anthropology's Method in a Time of Transition*. Ithaca, NY: Cornell University Press.

Feder, B. J. 1989. "Who Will Subscribe to the Valdez Principles?" *New York Times*, September 10.

Feit, H. A. 1987. "Waswanipi Cree Management of Land and Wildlife: Cree Cultural Ecology Revisited." In *Native Peoples: Native Lands*, ed. B. Cox, 75–91. Ottawa: Carleton University Press.

Fenge, T. 1999. "Humanitarian Assistance to People in Arctic Russia: A Special Role for Canada." In *Inuit in Global Issues*. Ottawa: ICC. www.inuitcircumpolar.com.

Fienup-Riordan, A. 1990. *Eskimo Essays*. New Brunswick, NJ: Rutgers University Press.

Fifka, M. S. 2011. "Corporate Responsibility Reporting and Its Determinants in Comparative Perspective—A Review of the Empirical Literature and a Meta-analysis." *Business Strategy and the Environment* 22, no. 1: 1–35.

Fischer, M. M. J. 2003. *Emergent Forms of Life and the Anthropological Voice*. Durham, NC: Duke University Press.

Fischer, M. M. J. 2009. *Anthropological Futures*. Durham, NC: Duke University Press.

Fisher-Vanden, K., and K. S. Thorburn. 2008. "Voluntary Corporate Environmental Initiatives and Shareholder Wealth." *Center for Economic Policy Research*. www.cepr.net.

Fisher, K. 1997. "Locating Frames in the Discursive Universe." *Sociological Research Online* 2, no. 3. www.socresonline.org.uk/2/3/4.html.

Fiske, J. 1996. *Media Matters: Race and Gender in U.S. Politics*. Minneapolis: University of Minnesota Press.

Fleck, L. [1935] 1981. *Genesis and Development of a Scientific Fact*. Chicago: University of Chicago Press.

Fora.tv. 2010. "Evangelical Richard Cizik Reflects on Ousting from NAE." www.youtube.com/watch?v=laj_zMG5834&feature=fvwk.

Fortun, K. 2001. *Advocacy after Bhopal*. Chicago: University of Chicago Press.

Fortun, K. 2003. "Ethnography in/of/as Open Systems." *Reviews in Anthropology* 32:171–90.

Fortun, K., and M. Fortun. 2005. "Scientific Imaginaries and Ethical Plateaus in Contemporary U.S. Toxicology." *American Anthropologist* 107:43–54.

Foucault, M. 1995. *Discipline and Punish: The Birth of the Prison*. New York: Vintage Books.

Foucault, M. 2003. *"Society Must Be Defended": Lectures at the Collège de France, 1975–1976*. New York: Picador.

Friedman, M. 1970. "The Social Responsibility of Business Is to Increase Its Profits." *New York Times Magazine*, September 13.

Fuss, D. 1989. *Essentially Speaking: Feminism, Nature, and Difference*. New York: Routledge.

Gamson, W. A. 1992. *Talking Politics*. Cambridge: Cambridge University Press.

Gans, H. 1979. *Deciding What's News: A Study of* CBS Evening News, NBC Nightly News, Newsweek, *and* Time. New York: Pantheon.

Gans, H. 2003. *Democracy and News*. Oxford: Oxford University Press.

Gelbspan, R. 2004. *Boiling Point: How Politicians, Big Oil and Coal, Journalists, and Activists Are Fueling the Climate Crisis—and What We Can Do to Avert Disaster*. New York: Basic Books.

Gelobter, M., M. K. Dorsey, L. Fields, T. Goldtooth, A. Mendiratta, R. Moore, R. Morello-Frosch, P. M. Shepard, and G. Torres. 2005. "The Soul of Environmentalism: Reconsidering Transformational Politics in the 21st Century." *Redefining Progress*. http://community-wealth.org/content/soul-environmentalism-reconsidering -transformational-politics-21st-century.

Gieryn, T. 1998. *Cultural Boundaries of Science: Credibility on the Line*. Chicago: University of Chicago Press.

Ginsberg, F., L. Abu-Lughod, and B. Larkin. 2002. *Media Worlds: Anthropology on New Terrain*. Berkeley: University of California Press.

Gitlin, T. 1980. *The Whole World Is Watching: Mass Media in the Making and Unmaking of the New Left*. Berkeley: University of California Press.

Gitlin, T. 1998. "Public Sphere or Public Sphericules." In *Media, Ritual and Identity*, ed. T. Liebes and J. Curran, 168–74. London: Routledge.

Gitlin, T. 2002. *Media Unlimited: How the Torrent of Images and Sounds Overwhlems Our Lives*. New York: Henry Holt.

Gladstone, B. 2006. "Dangerous Extremes." *On the Media*. NPR.

Glasser, T. L., and S. Craft. 1998. "Public Journalism and the Search for Democratic Ideals." In *Media, Ritual and Identity*, ed. T. Liebes and J. Curran, 203–18. New York: Routledge.

Goffman, E. 1974. *Frame Analysis: An Essay on the Organization of Experience*. New York: Harper and Row.

Goodstein, L. 2006. "Evangelical Leaders Join Global Warming Initiative." *New York Times*, February 8. www.nytimes.com/2006/02/08/national/08warm.html?page wanted=all.

Goodstein, L. 2007. "Evangelical's Focus on Climate Draws Fire of Christian Right." *New York Times*, March 3. www.nytimes.com/2007/03/03/us/03evangelical.html.

Goodstein, L. 2008. "Evangelical Lobbyist Resigns." *New York Times*, December 12. www.nytimes.com/2008/12/12/washington/12brfs-EVANGELICALL_BRF.html.

Gore, A. 2006. *An Inconvenient Truth: The Planetary Emergency of Global Warming and What We Can Do about It*. Emmaus, PA: Rodale.

Gore, A. 2011. "Climate of Denial." *Rolling Stone*, June 22. www.rollingstone.com /politics/news/climate-of-denial-20110622.

Grossman, C. L. 2007. "Evangelical: Can the 'E-word' Be Saved?" *USA Today*, January 23. http://usatoday30.usatoday.com/news/religion/2007–01–22-evangelicals -usat_x.htm?csp=1.

Grove, R. 1996. *Green Imperialism: Colonial Expansion, Tropical Island Edens and the Origins of Environmentalism, 1600–1860*. Cambridge: Cambridge University Press.

Gunther, M. 2007. "Dell Gets on the Environmental Bandwagon." *Fortune*, March 8. http://money.cnn.com/2007/03/08/magazines/fortune/pluggedin_gunther_dell recycle.fortune/index.htm?postversion=2007030809.

Habermas, J. 1962. *The Structural Transformation of the Public Sphere: An Inquiry into a Category of Bourgeois Society*. Cambridge: Polity Press.

Habermas, J. 1976. *Legitimation Crisis*. London: Heinemann Educational Books.

Harley, J. B. 1988. "Maps, Knowledge, and Power." In *The Iconography of Landscape: Essays on the Symbolic Representation, Design and Use of Past Environments*, ed. D. Cosgrove and S. Daniels, 277–312. Cambridge: Cambridge University Press.

Hannerz, U. 2004. *Foreign News: Exploring the World of Foreign Correspondents*. Chicago: University of Chicago Press.

Haraway, D. 1989. "Teddy Bear Patriarchy: Taxidermy in the Garden of Eden, New

York City, 1908–36." In *Primate Visions: Gender, Race, and Nature in the World of Modern Science*, 26–58. New York: Routledge.

Haraway, D. 1991. *Simians, Cyborgs, and Women: The Reinvention of Nature*. New York: Routledge.

Haraway, D. 1997. *Modest_Witness@Second_Millennium. FemaleMan©_Meets_Onco Mouse™: Feminism and Technoscience*. New York: Routledge.

Harding, S. 1991. "Representing Fundamentalism: The Problem of the Repugnant Cultural Other." *Social Research* 58:373–93.

Harding, S. 2000. *The Book of Jerry Falwell: Fundamentalist Language and Politics*. Princeton, NJ: Princeton University Press.

Hardt, M., and A. Negri. 2000. *Empire*. Cambridge, MA: Harvard University Press.

Hardt, M., and A. Negri. 2004. *Multitude: War and Democracy in the Age of Empire*. New York: Penguin Press.

Harvard Business Review. 2003. *Harvard Business Review on Corporate Social Responsibility*. Cambridge, MA: Harvard Business School.

Herman, E. S., and N. Chomsky. 1988. *Manufacturing Consent: The Political Economy of the Mass Media*. New York: Pantheon.

Hermida, A. 2011. "Fluid Spaces, Fluid Journalism: The Role of the Active Recipient." In *Participatory Journalism: Guarding Open Gates at Online Newspapers*, ed. J. B. Singer et al. Hoboken, NJ: Wiley-Blackwell.

Hilgartner, S. 2000. *Science on Stage: Expert Advice as Public Drama*. Palo Alto, CA: Stanford University Press.

Hilts, P. J. 2000. "Dioxin in Arctic Circle Is Traced to Sources Far to the South. *New York Times*, October 17. www.nytimes.com/2000/10/17/science/dioxin-in-arctic -circle-is-traced-to-sources-far-to-the-south.html.

Hinden, S. 1990. "Joan Bavaria's Crusade for the Environment." *Washington Post*, December 23.

Ho, K. 2009. *Liquidated: An Ethnography of Wall Street*. Durham, NC: Duke University Press.

Ho, K. 2010. "Outsmarting Risk: From Bonuses to Bailouts." *AnthroNow*, May 14. http:// anthronow.com/online-articles/outsmarting-risk-from-bonuses-to-bailouts.

Hoffman, A. 1996. "A Strategic Response to Investor Activism." *Sloan Management Review* 37:51–64.

Hoffman, M. J. 2011. *Climate Governance at the Crossroads: Experimenting with a Global Response after Kyoto*. Oxford: Oxford University Press.

Hoggan, J., and R. Littlemore. 2009. *Climate Cover-Up: The Crusade to Deny Global Warming*. Vancouver: Greystone Books.

Holdren, J., and P. Herrera. 1971. *Energy, a Crisis in Power*. San Francisco: Sierra Club.

Hulme, M. 2009. *Why We Disagree about Climate Change: Understanding Controversy, Inaction, and Opportunity*. Cambridge: Cambridge University Press.

Hunter, J. 2008. *A New Kind of Conservative*. Ventura, CA: Regal.

Ignatieff, M. 2001. *Human Rights as Politics and Idolatry*. Princeton, NJ: Princeton University Press.

Inglis, J. T., ed. 1993. *Traditional Ecological Knowledge: Concepts and Cases*. Ottawa:

International Program on Traditional Ecological Knowledge and International Development Research Centre.

Intergovernmental Panel on Climate Change. 2007. *IPCC Fourth Assessment Report: Climate Change 2007*. http://www.ipcc.ch/ipccreports/assessments-reports.htm.

Inuit Tapirisat of Canada. 1977. *Speaking for the First Citizens of the Canadian Arctic*. Ottawa: Inuit Tapirisat of Canada.

Irwin, A., and B. Wynne, eds. 2004. *Misunderstanding Science? The Public Reconstruction of Science and Technology*. Cambridge: Cambridge University Press.

James, S. D. 2008. "New Evangelists Buck the Christian Right." *ABC News*.

Jasanoff, S. 1990. *The Fifth Branch: Science Advisers as Policymakers*. Cambridge, MA: Harvard University Press.

Jasanoff, S. 1991. "Acceptable Evidence in a Pluralistic Society." In *Acceptable Evidence: Science and Values in Hazard Management*, ed. D. May and R. Hollander, 29–47. New York: Oxford University Press.

Jasanoff, S. 2003. "Breaking the Waves in Science Studies: Comment on H. M. Collins and Robert Evans, 'The Third Wave of Science Studies.'" *Social Studies of Science* 33, no. 3: 389–400.

Jasanoff, S. 2004. *States of Knowledge: The Co-production of Science and Social Order*. London: Routledge.

Jasanoff, S. 2005. *Designs on Nature: Science and Democracy in Europe and the United States*. Princeton, NJ: Princeton University Press.

Jasanoff, S. 2010. "A New Climate for Society." *Theory, Culture, and Society* 27, nos. 2–3: 233–53.

Jasanoff, S., and M. L. Martello, eds. 2004. *Earthly Politics: Local and Global in Environmental Governance*. Cambridge, MA: MIT Press.

Jenkins, H. 2006a. *Convergence Culture: Where Old and New Media Collide*. New York: New York University Press.

Jenkins, H. 2006b. *Fans, Bloggers, and Gamers: Exploring Participatory Culture*. New York: New York University Press.

Jenkins, H., S. Ford, and J. Green. 2012. *Spreadable Media: Creating Value and Meaning in a Networked Culture*. New York: New York University Press.

Jenkins, H., and D. Thorburn, eds. 2004. *Democracy and New Media*. Cambridge, MA: MIT Press.

Johnson, F. 2010. "SEC to Act Quickly on Risk Disclosure." *Wall Street Journal*, July 10. http://online.wsj.com/news/articles/SB10001424052748704075604575356842806735212.

Johnson, K. 2006. *E. O. Wilson's The Creation*. http://www.chestertonhouse.org/node/427.

Kay, P., and W. Kempton. 1984. "What Is the Sapir-Whorf Hypothesis?" *American Anthropologist* 86, no. 1: 65–79.

Keskitalo, E. C. H. 2004. *Negotiating the Arctic: The Construction of an International Region*. New York: Routledge.

Kintisch, E. 2006. "Evangelicals, Scientists Reach Common Ground on Climate Change." *Science* 311:1082–83.

Klinger, W. S. 1994. "Social Investing in a Changing World." *Best's Review*, February.

Kluger, J. 2005. "Are We Making Hurricanes Worse? The Impact of Global Warming." *Time*, October 3. http://content.time.com/time/nation/article/0,9263,760105100,00.html.

Kluger, J. 2006. "Kerry Emanuel." *Time*, May 8. http://content.time.com/time/specials/packages/article/0,28804,1975813_1975844_1976436,00.html.

Knorr-Cetina, K. 1999. *Epistemic Cultures: How Sciences Make Knowledge*. Cambridge, MA: Harvard University Press.

Koerner, L. 1999. *Linnaeus: Nature and Nation*. Cambridge, MA: Harvard University Press.

Kovach, B., and T. Rosenstiel. 2007. *The Elements of Journalism: What Newspeople Should Know and the Public Should Expect*. New York: Three Rivers Press.

Krech, S. 2000. *The Ecological Indian: Myth and History*. New York: W. W. Norton.

Krosnick, J. A., A. L. Holbrook, L. Lowe, and P. S. Visser. 2006. "The Origins and Consequences of Democratic Citizens' Policy Agendas: A Study of Popular Concern about Global Warming." *Climatic Change* 77, nos. 1–2: 7–43.

Kuhn, T. 1962. *The Structure of Scientific Revolutions*. Chicago: University of Chicago Press.

LaBarre, S. 2013. "Why We're Shutting Off Our Comments." *Popular Science*, September 24. www.popsci.com/science/article/2013-09/why-were-shutting-our-comments.

LaDuke, W. 1999. *All Our Relations: Native Struggles for Land and Life*. Cambridge, MA: South End Press.

Lahsen, M. 1998. "Climate Rhetoric: Constructions of Climate Science in the Age of Environmentalism." PhD diss., Rice University, Houston.

Lahsen, M. 2005a. "Seductive Simulations? Uncertainty Distribution around Climate Models." *Social Studies of Science* 35, no. 6: 895–922.

Lahsen, M. 2005b. "Technocracy, Democracy, and U.S. Climate Politics: The Need for Demarcations." *Science, Technology, and Human Values* 30:137–69.

Lahsen, M. 2008. "Experiences of Modernity in the Greenhouse: A Cultural Analysis of a Physicist 'Trio' Supporting the Backlash against Global Warming." *Global Environmental Change* 18:204–19.

Lahsen, M. 2010. "The Social Status of Climate Change Knowledge: An Editorial Essay." *Wiley Interdisciplinary Reviews: Climate Change* 1:162–71.

Lakoff, G. 2004. *Don't Think of an Elephant: Know Your Values and Frame the Debate—The Essential Guide for Progressives*. White River Junction, VT: Chelsea Green Publishing.

Lalonde, A. 1993. "African Indigenous Knowledge and Its Relevance to Sustainable Development." In *Traditional Ecological Knowledge: Concepts and Cases*, ed. J. T. Inglis. Ottawa: Canadian Museum of Nature and International Development Research Centre.

Latour, B. 1988. *Science in Action: How to Follow Scientists and Engineers through Society*. Cambridge, MA: Harvard University Press.

Latour, B. 1991. *We Have Never Been Modern*. Cambridge, MA: Harvard University Press.

Latour, B. 2004a. *Politics of Nature: How to Bring the Sciences into Democracy*. Cambridge, MA: Harvard University Press.

Latour, B. 2004b. "Why Has Critique Run Out of Steam? From Matters of Fact to Matters of Concern." *Critical Inquiry* 30, no. 2: 225–48.

Latour, B. 2005. *Reassembling the Social: An Introduction to Actor-Network-Theory*. Oxford: Oxford University Press.

Latour, B., and S. Woolgar. 1979. *Laboratory Life: The Construction of Scientific Facts*. London: Sage.

Lee, J. 2003. "A Call for Softer, Greener Language." *New York Times*, March 2. www .nytimes.com/2003/03/02/us/a-call-for-softer-greener-language.html.

Leiserowitz, A., E. Maibach, and C. Roser-Renouf. 2008. "Global Warming's 'Six Americas': An Audience Segmentation." Fairfax, VA: George Mason University Center for Climate Change Communication.

Leiserowitz, A. A. 2004. "Before and After *The Day after Tomorrow*: A U.S. Study of Climate Change Risk Perception." *Environment* 46, no. 9: 22–37.

Lévi-Strauss, C. 1966. *The Savage Mind*. Chicago: University of Chicago Press.

Levy, P. 1997. *Collective Intelligence: Mankind's Emerging World in Cyberspace*. New York: Perseus Books.

Lindisfarne, N. 2010. "Cochabamba and Climate Anthropology." *Anthropology Today* 26:1–3.

Lippmann, W. 1922. *Public Opinion*. New York: Free Press.

Little, A. 2005. "Cizik Matters: An Interview with Green Evangelical Leader Richard Cizik." Grist.org, October 6. http://grist.org/article/cizik/.

Loukacheva, N. 2007. *The Arctic Promise: Legal and Political Autonomy of Greenland and Nunavut*. Toronto: University of Toronto Press.

Lowe, T. 2006. "Is This Climate Porn? How Does Climate Change Communication Affect Our Perceptions and Behaviour?," Tyndall Centre Working Paper 98. Norwich, UK: University of East Anglia.

Lynch, M. 1998. "The Discursive Production of Uncertainty." *Social Studies of Science* 28:829–67.

Lynge, A. 1993. *Inuit: The Story of the Inuit Circumpolar Conference*. Nuuk: ICC.

Lynge, F. 1992. *Arctic Wars, Animal Rights, Endangered Peoples*. Hanover, NH: Dartmouth College and University Press of New England.

Maaka, R. C., and C. Andersen, eds. 2006. *The Indigenous Experience: Global Perspectives*. Toronto: Canadian Scholars' Press.

Malaurie, J. 2007. *Hummocks: Journeys and Inquires among the Canadian Inuit*. Montreal: McGill-Queen's University Press.

Mann, M. E. 2012. *The Hockey Stick and the Climate Wars*. New York: Columbia University Press.

Marcus, G., ed. 1995. *Technoscientific Imaginaries: Conversations, Profiles, and Memoirs*. Chicago: University of Chicago Press.

Marcus, G. 1998. *Ethnography through Thick and Thin*. Princeton, NJ: Princeton University Press.

Marcus, G., ed. 2000. *Para-Sites: A Casebook against Cynical Reason*. Chicago: University of Chicago Press.

Marcus, G., and M. M. J. Fischer. 1999. *Anthropology as Cultural Critique: An Experimental Moment in the Human Sciences*. Chicago: University of Chicago Press.

Martin, J. D. 2011. "What's So Wrong with 'Parachute Journalism'?" *Columbia Journalism Review*, May 26. www.cjr.org/behind_the_news/whats_so_wrong_with _parachute.php.

Martello, M. L. 2008. "Arctic Indigenous Peoples as Representations and Representatives of Climate Change." *Social Studies of Science* 38:351–76.

Mason, A. 2002. "The Rise of an Alaskan Native Bourgeoisie." *Étude/Inuit/Studies* 26:5–22.

May, S., G. Cheney, and J. Roper, eds. 2007. *The Debate over Corporate Social Responsibility*. Oxford: Oxford University Press.

Mazur, A., and J. Lee. 1993. "Sounding the Global Alarm: Environmental Issues in the US National News." *Social Studies of Science* 23, no. 4: 681–720.

McBride, K., and T. Rosenstiel. 2013. *The New Ethics of Journalism: Principles for the 21st Century*. Thousand Oaks, CA: CQ Press.

McChesney, R. W. 1999. *Rich Media, Poor Democracy: Communication Politics in Dubious Times*. New York: New Press.

McKenna, G. 2007. *The Puritan Origins of American Patriotism*. New Haven, CT: Yale University Press.

McQuaid, J., and M. Schleifstein. 2006. *Path of Destruction: The Devastation of New Orleans and the Coming Age of Superstorms*. New York: Little, Brown.

Melucci, A. 1989. *Nomads of the Present: Social Movements and Individual Needs in Contemporary Society*. London: Hutchison Radius.

Melucci, A. 1996. *Challenging Codes: Collective Action in an Information Age*. Cambridge: Cambridge University Press.

Merritt, D. 1995. *Public Journalism and Public Life*. Mahwah, NJ: Lawrence Erlbaum.

Merton, R. 1973. *The Sociology of Science: Theoretical and Empirical Investigations*. Chicago: Chicago University Press.

Miller, C. A., and P. N. Edwards, eds. 2001. *Changing the Atmosphere: Expert Knowledge and Environmental Governance*. Cambridge, MA: MIT Press.

Miller, P. 2000. "Tracking Dioxins to the Arctic." Commission for Environmental Cooperation. www.cec.org/Page.asp?PageID=122&ContentID=2416.

Mills, A. 1994. *Eagle Down Is Our Law: Witsuwit'en Law, Feasts, and Land Claims*. Vancouver: UBC Press.

Mitchell, D. C. 1997. *Sold American: The Story of Alaska Natives and Their Land, 1867–1959, the Army to Statehood*. Hanover, NH: University Press of New England.

Mitchell, M. 1996. *From Talking Chiefs to a Native Corporate Elite: The Birth of Class and Nationalism among Canadian Inuit*. Montreal: McGill-Queen's University Press.

Monet, D., and A. Wilson. 1992. *Colonialism on Trial: Indigenous Land Rights and the Gitksan and Wet'suwet'en Sovereignty Case*. Philadelphia: New Society.

Mooney, C. 2007. *Storm World: Hurricanes, Politics, and the Battle over Global Warming*. Orlando: Harcourt.

Moser, S. C., and L. Dilling, eds. 2007. *Creating a Climate for Change: Communicating Climate Change and Facilitating Social Change*. Cambridge: Cambridge University Press.

Moyers, B. 2006. "Is God Green?" *Moyers on America*. PBS, October 9.

Munson, E. S., and C. A. Warren, eds. 1997. *James Carey: A Critical Reader*. Minneapolis: University of Minnesota Press.

Nadasdy, P. 1999. "The Politics of TEK: Power and the 'Integration' of Knowledge." *Arctic Anthropology*, 36, nos. 1–2: 1–18.

Nadasdy, P. 2003. *Hunters and Bureaucrats: Power, Knowledge, and Aboriginal-State Relations in the Southwest Yukon*. Vancouver: UBC Press.

National Association of Evangelicals. 2008a. *Revision*. No longer available on website.

National Association of Evangelicals. 2008b. "Richard Cizik Resigns from NAE," December 11. http://www.nae.net.

Neff, D. 2009a. "Can We Separate Creation Care from Political Action?" *Christianity Today LiveBlog*, April 24. www.christianitytoday.com/gleanings/2009/april/can-we-separate-creation-care-from-political-action.html.

Neff, D. 2009b. "Creation Care without the Baggage." *Christianity Today LiveBlog*, May 17. www.christianitytoday.com/gleanings/2009/may/creation-care-without-baggage.html.

Niezen, R. 2003. *The Origins of Indigenism: Human Rights and Politics of Identity*. Berkeley: University of California Press.

Nikiforuk, A. 2010. *Tar Sands: Dirty Oil and the Future of a Continent*. Vancouver: Greystone Books.

Nisbet, M. C. 2009. "Communicating Climate Change: Why Frames Matter for Public Engagement." *Environment: Science and Policy for Sustainable Development* 51, no. 2: 12–23.

Nisbet, M., and D. A. Scheufele. 2009. "What's Next for Science Communication? Promising Directions and Lingering Distractions." *American Journal of Botany* 96:1767–78.

Nisbet, M. C., and C. Mooney. 2007. "Framing Science." *Science* 316, no. 5821: 56.

Nisbet, M. C., and T. Myers. 2007. "The Polls—Trends Twenty Years of Public Opinion about Global Warming." *Public Opinion Quarterly* 71, no. 3: 444–70.

Nowotny, H., P. Scott, and M. Gibbons. 2001. *Re-thinking Science: Knowledge and the Public in an Age of Uncertainty*. Malden, MA: Blackwell.

NPR. 2008. "Rev. Richard Cizik on God and Global Warming." *Fresh Air*, December 2.

Oozeva, C., C. Noongwook, G. Noongwook, C. Alowa, and I. Krupnik. 2004. *Watching Ice and Weather Our Way: Sikumengllu Eslamengllu Esghapalle-Ghput*. Washington, DC, and Savoonga, AK: Arctic Studies Center, Smithsonian Institution and Savoonga Whaling Captains Association.

Oreskes, N. 2004a. "Beyond the Ivory Tower: The Scientific Consensus on Climate Change." *Science* 306, no. 5702: 1686.

Oreskes, N. 2004b. "Science and Public Policy: What's Proof Got to Do with It?" *Environmental Science and Policy* 7:369–83.

Oreskes, N. 2004c. "Undeniable Global Warming." *Washington Post*, December 26. www.washingtonpost.com/wp-dyn/articles/A26065-2004Dec25.html.

Oreskes, N. 2003. "The Role of Quantitative Models in Science." In *Models in Ecosystem Science*, ed. C. D. Canham, J. J. Cole, and W. K. Lauenroth, 14–31. Princeton, NJ: Princeton University Press.

Oreskes, N., and E. Conway. 2010. *Merchants of Doubt: How a Handful of Scientists Obscured the Truth on Issues from Tobacco Smoke to Global Warming.* New York: Bloomsbury.

Palen, J. 1999. "Objectivity as Independence: Creating the Society of Environmental Journalists, 1989–1997." *Science Communication* 21, no. 2: 156–71.

Palin, S. 2008. "Op-Ed Contributor: Bearing Up." *New York Times*, January 5. www.nytimes.com/2008/01/05/opinion/05palin.html.

PBS Nova. 2007. "Judgment Day: Intelligent Design on Trial." *Nova*, November 13. www.pbs.org/wgbh/nova/evolution/intelligent-design-trial.html.

Peters, C., and M. J. Broersma, eds. 2013. *Rethinking Journalism: Trust and Participation in a Transformed News Landscape.* New York: Routledge.

Pew Research Center for the People and the Press. 2008. "A Deeper Partisan Divide over Global Warming," news release, May 8. http://people-press.netcampaign.com/reports/.

Pew Research Religion and Public Life Project. 2007. "The Religious Factor in the 2008 Election." Pew Forum Faith Angle Conference, Key West, December 4. www.pewforum/2007/12/04/the-religion-factor-in-the-2008-election/.

Pielke, R. A. 2007. *The Honest Broker: Making Sense of Science in Policy and Politics.* Cambridge: Cambridge University Press.

Pollock, A. 2012. *Medicating Race: Heart Disease and Durable Preoccupations with Difference.* Durham, NC: Duke University Press.

Porter, T. M. 1992. "Quantification and the Accounting Ideal in Science." *Social Studies of Science* 22, no. 4: 633–51.

Povinelli, E. A. 2001. "Radical Worlds: The Anthropology of Incommensurability and Inconceivability." *Annual Review of Anthropology* 30:319–34.

Rabinow, P. 1992. "Artificiality and Enlightenment: From Sociobiology to Biosociality." In *The Science Studies Reader*, ed. M. Biagioli, 407–16. New York: Routledge.

Rabinow, P. 2007. *Marking Time: On the Anthropology of the Contemporary.* Princeton, NJ: Princteon University Press.

Rahmstorf, S., M. Mann, R. Benestad, G. Schmidt, and W. Connolley. 2005. "Hurricanes and Global Warming—Is There a Connection?" *RealClimate*, September. www.realclimate.org/index.php/archives/2005/09/hurricanes-and-global-warming/.

Rapp, Rayna. 2000. *Testing Women, Testing the Fetus: The Social Impact of Amniocentesis in America.* New York: Routledge.

Rensberger, B. 1992. "As Earth Summit Nears, Consensus Still Lacking on Global Warming's Cause." *Washington Post*, May 31.

Revkin, A. 2008. "Hurricane Expert Reassesses Link to Warming." *New York Times*, April 12.

Revkin, A. 2010. "Dot Earth 2.0." *New York Times*, March 31. http://dotearth.blogs .nytimes.com/2010/03/31/dot-earth-2-0/.

Rheingold, H. 2003. *Smart Mobs: The Next Social Revolution*. New York: Basic Books.

Roberts, D. 2006a. "Rev. Joel Hunter Speaks Out on Broadening the Evangelical Agenda." *Grist.org*, December 21. http://grist.org/article/hunter/.

Roberts, D. 2006b. "The Soul of DeWitt: An Interview with Environmental Scientist and Evangelical Leader Calvin DeWitt." *Grist.org*, October 17. http://grist.org /article/dewitt/.

Roberts, W. L. 1993. "Sun's Bold Strategy for Environment." *Philadelphia Business Journal*.

Robinson, T. 2008. *Small Footprint, Big Handprint: How to Live Simply and Love Extravagantly*. Boise, ID: Ampelon Publishing.

Robinson, T., and J. Chatraw. 2006. *Saving God's Green Earth: Rediscovering the Church's Responsibility to Environmental Stewardship*. Norcross, GA: Ampelon Publishing.

Rosen, J. 1999. *What Are Journalists For?* New Haven, CT: Yale University Press.

Rouse, J. 1992. "What Are Cultural Studies of Scientific Knowledge?" *Configurations* 1:57–94.

Rudiak-Gould, P. 2011. "Climate Change and Anthropology: The Importance of Reception Studies." *Anthropology Today* 27, no. 2: 9–12.

Russell, C. 2008. "Climate Change: Now What?" *Columbia Journalism Review*, July 8. http://www.cjr.org/feature/climate_change_now_what.php?page=all.

Russell, C. 2009a. "*Globe* Kills Health/Science Section, Keeps Staff." *Columbia Journalism Review*, March 4. http://www.cjr.org/the_observatory/globe_kills_health science_sect.php?page=all.

Russell, C. 2009b. "*Washington Post* Pools Its Resources." *Columbia Journalism Review*, March 6. http://www.cjr.org/the_observatory/washington_post_pools_its_reso .php?page=all&print=true.

Salter, C. 2006. "Moving Heaven and Earth." *Fast Company*, June 1. http://www .fastcompany.com/56858/moving-heaven-and-earth.

Salter, C. 2008. "Evangelical Environmentalist Richard Cizik Forced Out of the NAE." *Fast Company*, December 23. http://www.fastcompany.com/1121508/evangelical -environmentalist-richard-cizik-forced-out-nae.

Sambrook, R. 2011. *Are Foreign Correspondents Redundant?* Oxford: University of Oxford.

Sanal, A. 2011. *New Organs within Us: Transplants and the Moral Economy*. Durham, NC: Duke University Press.

Sanyal, R. N., and J. S. Neves. 1991. "The Valdez Principles: Implications for Corporate Social Responsibilty." *Journal of Business Ethics* 10:883–90.

Sataline, S. 2008. "For Some Evangelicals, GOP Ties Are No Longer Binding." *Wall Street Journal*, August 22. http://online.wsj.com/news/articles/SB121937082033962551.

Saussure, F. de. 1986. *Course in General Linguistics*, ed. C. Bally, A. Sechehaye, and A. Riedlinger. Peru, IL: Open Court.

Sauvageau, F. 2012. "The Uncertain Future of the News." In *How Canadians Communicate IV: Media and Politics*, ed. D. Taras and C. Waddell. Edmonton: Athabasca University Press.

Sayre, C. 2006. "The Year in Buzzwords 2006." *Time*, December 17.

Schagen, S. 2007. "Live Earth: How Green Are the Concerts for a Climate in Crisis?" *Grist.org*, July 3.

Scheufele, D. A. 1999. "Framing as a Theory of Media Effects." *Journal of Communication* 49:103–22.

Schneider, S. H. 2011. "Mediarology." In *Understanding and Solving the Climate Change Problem*. http://stephenschneider.stanford.edu/Mediarology/Mediarology.html.

Schrag, D. 2006. "On a Swift Boat to a Warmer World." *Boston Globe*, December 17. http://www.boston.com/news/globe/editorial_opinion/oped/articles/2006/12/17/on_a_swift_boat_to_a_warmer_world/.

Schudson, M. 1978. *Discovering the News: A Social History of American Newspapers*. New York: Basic Books.

Schudson, M. 1995. *The Power of News*. Cambridge, MA: Harvard University Press.

Schudson, M. 1998. *The Good Citizen: A History of American Civic Life*. New York: Free Press.

Schudson, M. 2001. "The Objectivity Norm in American Journalism." *Journalism* 2, no. 2: 149–70.

Schudson, M. 2002. *The Sociology of News*. New York: W. W. Norton.

Schudson, M. 2005. "Autonomy from What?" In *Bourdieu and the Journalistic Field*, ed. Rodney Benson. Cambridge: Polity, 214–23.

Schudson, M. 2013. "Would Journalism Please Hold Still!" In *Rethinking Journalism: Trust and Participation in a Transformed News Landscape*, edited by C. Peters and M. Broersma, 191–99. New York: Routledge.

Scruggs, L., and S. Benegal. 2012. "Declining Public Concern about Climate Change: Can We Blame the Great Recession?" *Global Environmental Change* 22:505–15.

Seife, C. 2010. *Proofiness: How You're Being Fooled by the Numbers*. New York: Penguin.

Semeniuk, R. 2007. *Among the Inuit*. Vancouver: Raincoast Books.

Shapin, S. 1995. "Cordelia's Love: Credibility and the Social Studies of Science." *Perspectives on Science* 3:255–75.

Shapin, S., and S. Schaffer. 1989. *Leviathan and the Air-Pump: Hobbes, Boyle, and the Experimental Life*. Princeton, NJ: Princeton University Press.

Shellenberger, M., and T. Nordhaus. 2005. "The Death of Environmentalism: Global Warming Politics in a Post-environmental World." Environmental Grantmakers Association. http://www.grist.org/news/maindish/2005/01/13/doe-reprint.

Simon, M. 1996. *Inuit: One Future—One Arctic*. Peterborough, ON: Cider Press.

Singer, J. B. 2005. "The Political J-blogger: 'Normalizing' a New Media Form to Fit Old Norms and Practices." *Journalism* 6:173.

Singer, J. B. 2007. "Contested Autonomy: Professional and Popular Claims on Journalistic Norms." *Journalism Studies* 8, no. 1: 79–95.

Singer, J. B., A. Hermida, D. Domingo, A. Heinonen, S. Paulussen, T. Quandt, Z. Reich, and M. Vujnovic. 2011. *Participatory Journalism: Guarding Open Gates at Online Newspapers*. Hoboken, NJ: Wiley-Blackwell.

Sleeth, M. 2007. *Serve God and Save the Planet: A Christian Call to Action*. Grand Rapids, MI: Zondervan.

Snow, D. A., E. B. Rochford, S. K. Worden, and R. D. Benford. 1986. "Frame Alignment Processes, Micromobilization and Movement Participation." *American Sociological Review* 51:464–81.

Snow, D. A., and R. D. Benford. 1988. "Ideology, Frame Resonance and Participant Mobilization." *International Social Movement Research* 1:197–219.

SourceWatch. 2010. *Greenwashing*. http://www.sourcewatch.org/index.php?title =Greenwashing.

Speth, J. G. 2004. *Red Sky at Morning: America and the Crisis of the Global Environment*. New Haven, CT: Yale University Press.

Spivak, G. 1990. *The Post-Colonial Critic: Interviews, Strategies, Dialogues*. New York: Routledge.

Steinberg. 1999. "The Talk and Back Talk of Collective Action: A Dialogic Analysis of Repertoires of Discourse among Nineteenth-Century English Cotton Spinners." *American Journal of Sociology* 105:736–80.

Stern, N. 2007. *The Economics of Climate Change: The Stern Review*. Cambridge: Cambridge University Press.

Sunder-Rajan, K. 2006. *Biocapital: The Constitution of Postgenomic Life*. Durham, NC: Duke University Press.

Sze, J. 2006. "Boundaries and Border Wars: DES, Technology, and Environmental Justice." *American Quarterly* 58, no. 3: 791–814.

Terdiman, R. 1990. *Discourse/Counter-Discourse: The Theory and Practice of Symbolic Resistance in Nineteenth-Century France*. Ithaca, NY: Cornell University Press.

Traweek, S. 1988. *Beamtimes and Lifetimes: The World of High Energy Physicists*. Cambridge, MA: Harvard University Press.

Tsing, A. 2005. *Friction: An Ethnography of Global Connection*. Princeton, NJ: Princeton University Press.

Tumber, H. 2001. "Democracy in the Information Age: The Role of the Fourth Estate in Cyberspace." *Information, Communication and Society* 4:95–112.

Turkle, S. 1995. *Life on the Screen: Identity in the Age of the Internet*. New York: Touchstone.

Turner, F. 2006. *From Counterculture to Cyberculture: Stewart Brand, the Whole Earth Network, and the Rise of Digital Utopianism*. Chicago: University of Chicago Press.

Ungar, S. 1992. "The Rise and (Relative) Decline of Global Warming as a Social Problem." *Sociological Quarterly* 33:483.

Usher, N. 2010. "Why Spreadable Doesn't Equal Viral: A Conversation with Henry Jenkins." *Nieman Journalism Lab*, November 23.

U.S. Senate Committee on Environment and Public Works. 2006. "Examining Climate Change and the Media." Washington, DC: U.S. Government Printing Office.

Vakhtin, N. 1994. "Native Peoples of the Russian Far North." In *Polar Peoples: Self-Determination and Development*. London: Minority Rights.

van der Sluijs, J., J. van Eijndhoven, S. Shackley, and B. Wynne. 1998. "Anchoring Devices in Science for Policy: The Case of Consensus around Climate Sensitivity." *Social Studies of Science* 28:291–323.

Vu, M. 2007. "Evangelical's Global Warming Stance Disturbs Some Christian Leaders." *Christian Post*, March 6. http://www.christianpost.com/news/evangelical -s-global-warming-stance-disturbs-some-christian-leaders-26161/.

Vurro, C., and F. Perrini. 2011. "Making the Most of Corporate Social Responsibility Reporting: Disclosure Structure and Its Impact on Performance." *Corporate Governance* 11, no. 4: 459–74.

Wald, M. L. 1993. "Company News: Corporate Green Warrior; Sun Oil Takes Environmental Pledge." *New York Times*, February 11.

Walley, C. 2004. *Rough Waters: Nature and Development in an East African Marine Park*. Princeton, NJ: Princeton University Press.

Wallis, J. 2006. *God's Politics: Why the Right Gets It Wrong and the Left Doesn't Get It*. New York: HarperCollins.

Ward, B. 2008. *Communicating on Climate Change: An Essential Resource for Journalists, Scientists, and Educators*. Narragansett, RI: Metcalf Institute for Marine and Environmental Reporting.

Ward, S. J. A. 2004. *Invention of Journalism Ethics: The Path to Objectivity and Beyond*. Montreal: McGill-Queens University Press.

Ward, S. J. A. 2009. *Global Journalism Ethics*. Montreal: McGill-Queens University Press.

Warner, M. 1990. *The Letters of the Republic: Publication and the Public Sphere in Eighteenth-Century America*. Cambridge, MA: Harvard University Press.

Watt-Cloutier, S. 2007. "Nunavut Must Think Big, Not Small, on Polar Bears." *Nunatsiaq News*, February 1. http://www.nunatsiaq.com/test/archives/2007/701/70126 /opinionEditorial/opinions.html.

Weart, S. 2009. *The Discovery of Global Warming: The Public and Climate Change (cont.—since 1980)*. http://www.aip.org/history/climate/public2.htm#L_M011.

Weber, M., D. S. Owen, T. B. Strong, and R. Livingstone. 2004. *The Vocation Lectures: Science as a Vocation, Politics as a Vocation*. Indianapolis: Hackett.

Whorf, B. L. *Language, Thought and Reality*, ed. J. B. Carroll. Cambridge, MA: MIT Press.

Wildmon, D., T. Perkins, J. Dobson, G. L. Bauer, P. Weyrich et al. 2007. "Letter Regarding Richard Cizik and Global Warming," ed. L. Roy Taylor, chairman of the board, National Association of Evangelicals, Lawrenceville, GA.

Wilkinson, K. K. 2012. *Between God and Green: How Evangelicals Are Cultivating a Middle Ground on Climate Change*. Oxford: Oxford University Press.

Wilson, E. O. 2006. *The Creation: An Appeal to Save Life on Earth*. New York: W. W. Norton.

Winston, B. 1998. *Media Technology and Society: A History*. New York: Routledge.

Wittgenstein, L. [1953] 2001. *Philosophical Investigations*. Malden, MA: Blackwell.

Wittgenstein, L. [1969] 2008. *On Certainty*. Malden, MA: Blackwell.

Wohlforth, C. 2004. *The Whale and the Supercomputer: On the Northern Front of Climate Change*. New York: North Point Press.

Wynne, B. 2008. "Elephants in the Rooms Where Publics Encounter 'Science'?: A Response to Darrin Durant, 'Accounting for Expertise: Wynne and the Autonomy of the Lay Public.'" *Public Understanding of Science* 17, no. 1: 21–33.

Xanthaki, A. 2004. "Indigenous Rights in the Russian Federation: The Case of Numerically Small Peoples of the Russian North, Siberia, and Far East." *Human Rights Quarterly* 26:74–105.

Young, Oran. 1992. *Arctic Politics: Conflict and Cooperation in the Circumpolar North.* Hanover, NH: University Press of New England.

Index

Alvanna-Stimpfle, Megan, 43–44, 47, 267n6
American Electric Power (AEP), 217–19, 220
American Federation of Labor and Congress of Industrial Organizations (AFL-CIO), 279n6
American Geophysical Union (AGU), 187
Anaya, James, 65
Anchorage Declaration, 259–60, 269n18
Anderson, Alun, 76
Anderson, Benedict, 26, 266n15
anthropology of science scholarship, 2, 30; contextualization in, 34–36; multisited ethnography in, 5, 7, 31–36; parachute forms of, 33–34, 36; para-sites of, 281n1
anti-apartheid movement, 207, 209–13, 232, 239
anti-pollution activism, 207, 213, 239
Arbogast, Tod, 217
Arctic: focus on vulnerability of, 55, 267n8; indigenous groups of, 77; persistent organic pollutants (POPs) in, 67–68, 70, 72, 269n20; resource potentialities of, 74–78, 267n8, 271n11; sea ice melt and sea level rise in, 44–45, 46, 54, 75, 82, 99–100; as transnational region, 76–77; visible evidence of climate change in, 99–101. *See also* Inuit; Inuit Circumpolar Council; Iñupiat
Arctic Climate Impact Assessment (ACIA), 8, 54–55, 77, 167, 254, 276n3
Arctic Council, 77, 253
Arctic Energy Summit (AES), 75–76
Arctic indigenous peoples. *See* Inuit; Inuit Circumpolar Council; Iñupiat
Arctic Science Summit Week (ASSW), 43, 57–64, 258; indigenous participation in, 78; *Thin Ice* exhibit of, 58–60, 66–67, 74, 75, 268n12
A Rocha: Christians in Conservation, 275n19
articulation theory, 270n21

Au Sable Institute, 131, 144, 254, 275n19
Axworthy, Lloyd, 65

"Balance as Bias" (Boykoff and Boykoff), 69, 88, 94, 98, 193, 255
Ball, Jim, 131–33, 136–39, 141–49, 153–56, 160–61, 255, 275n18
Bank of America, 211
Bavaria, Joan, 209, 215, 279nn3–4, 279n7
Bavaria Awards for Building Sustainability into Capital Markets, 215
Bear Stearns, 235–36, 280n19
Beck, Ulrich, 115–16, 118, 170, 233, 236, 239, 250–51
Benestad, Rasmus, 96–98
Benkler, Yochai, 266n15
Bernal-Polanyi debates, 265n12
Besley, John C., 277n12
Between God and Green (Wilkinson), 273n4
Blair, Tony, 162, 257
blessing the facts, 126–27, 137–38, 146–47, 154–55, 160
blogosphere, 111–13, 143, 192–93, 272n15. *See also* social media
Bloor, David, 267n4
Boczkowski, Pablo, 24, 271n14
The Book of Jerry Falwell (Harding), 125
Boroditsky, Lera, 264n9
boundary-ordering devices, 169–70, 197
Bourdieu, Pierre, 30
Boxer, Barbara, 173, 176
Boykoff, Jules, 69, 88, 93–94, 98, 193, 255
Boykoff, Maxwell, 4, 69, 85–86, 88, 93–94, 98, 193, 255
British Columbia, 268n17
Brody, Hugh, 268n14
Buse, Uwe, 177–78
Bush, George W.: climate change policies of, 3, 69, 132, 141, 224, 254, 258; evangelical subculture under, 148
business community. *See* Ceres
Butcher, Jim, 230–31

California pension funds, 209
Campolo, Tony, 155

communication on climate change, 1–5; adjudication of expertise and, 37, 83, 98–99, 105, 125, 248; alternative discourses in, 6–10, 29–36; cliché in, 50–51; linking of moral meaning with science in, 70–74; methodology of analysis of, 10–15; presumed impartiality and credibility of, 2, 3, 18–20, 23; translation and relationship building with facts in, 13–15, 29–30, 50–52, 244, 264n9. *See also* journalism on climate change; vernaculars of meaning, ethics, and morality

Connolley, William, 96–98

Conrad, Kevin, 281n1

contextualization, 33–36

Conway, Eric, 4, 15, 88–89

coproduction of scientific knowledge, 5, 161, 172–73, 197, 264n3, 278n15

Cornwall Declaration, 132, 145–46, 257, 274n15

corporate social responsibility (CSR), 7–8, 206–7; greenwashing and, 220–23; shareholder resolutions on, 211–13, 217–19, 229, 239; stakeholder meetings on, 216–20, 239, 279n6, 279n9; sustainability reporting of, 204, 218–19, 223, 239. *See also* Ceres

Correll, Robert, 65, 74–75

cost of carbon, 227–28, 232, 279n12

Creation Care, 7, 37, 121–61, 249, 279n5; abortion debates and, 129, 135, 141, 152, 154–55, 157, 159; on climate science, 24, 123–25, 137–38, 147, 154–55; Evangelical Climate Initiative and, 6, 131–34, 137, 140–43, 157, 255, 257, 274n13, 274n15; leaders and messengers of, 9, 22, 125–31, 136–40, 146–47, 259; origins and target audience of, 134–36, 255, 274n11; skepticism and opposition to, 123, 131, 132, 136, 160; stewardship focus of, 123–24, 126, 129–30, 152, 155; translation and mobilization challenges of, 15, 29, 105, 125–31, 154, 159–61, 275n19; vernac-

ular framework of, 123–27, 140–42, 154–59, 247

Crichton, Michael, 272n19

Cronkite, Walter, 18–19

Cruikshank, Julie, 56–57, 73–74, 268n9

Curry, Judith, 97–99, 102, 149

Curwood, Steve, 203

The Daily Show, 272n17

Daly, Patricia, 213

The Day after Tomorrow, 174

Dean, Cornelia, 103–7, 193–95, 271n13, 277n13; on advocacy groups, 272n15; on journalist advocacy, 106–7; on journalist skepticism, 248; on science expert advocacy, 189–91

Dean, Jodie, 266n15

"The Death of Environmentalism" (Shellenberger and Nordhaus), 9, 255

definitions of climate change, 12–13, 51–52, 168, 245–47. *See also* translation

Deleuze, Gilles, 50

Delgamuukw v. British Columbia decision, 268n17

Dell, Inc., 217, 219, 279n12

Demming, David, 173

democracy: journalism's public education role in, 4–6, 10, 19–20, 22–23, 25–30, 91, 101–6, 116–17, 118, 266nn16–17; negotiation of complexity in, 251–52; responsibilities of citizenship in, 9, 20, 27, 30, 171, 191, 244, 251, 266n16

Dewey, John, 27–28, 276n7

DeWitt, Calvin, 131–32, 149, 152–53, 155, 254, 255

"Disruptive Environments" conference, 264n4, 281n1

Divine Wind (Emanuel), 184

Dobson, James, 129, 132, 145, 157, 257

Doppelt, Bob, 81–82, 89

Dot Earth blog, 112, 114, 117

double binds, 2–3, 5, 32, 35, 84, 248, 263n1, 278n15

Dover School District, Kitzmiller v., 126, 148

Dumit, Joseph, 11, 24–25, 276n6
Durkheim, Émile, 107–8

Earthjustice, 68–69
Earth Summit, 91, 144
Earth-Wise: A Biblical Response to Environmental Issues (DeWitt), 152–53
economic factors. *See* risk
economic recession of 2008, 234–36, 280n19
The Economics of Climate Change: The Stern Review, 162–68, 171–72, 178, 182, 230, 236, 257, 275nn1–2
Edwards, Paul, 4, 167
effectiveness in journalism, 90–94
Ehrlich, Paul, 181
Emanuel, Kerry, 16, 281n1; on coastal development, 237; *Nature* article on hurricanes, 16, 96, 97, 99, 235, 256; public engagement of, 184–89, 199; on scientists and the media, 193–94
Energy: A Crisis in Power (Holdren), 182
Enron bankruptcy scandal, 224–25, 235
enunciatory communities, 7, 248–49, 264n5. *See also* alternative discourses
Environmental Law Institute, 87
epistemological knowledge, 46–47, 52
epistemology, 46–47, 52; alarmism and, 87–89; differences in, 29, 245; negotiation with, 126; science and, 151, 165–66
Ethical Corporation, 237
ethnography, 32–36. *See also* multisited ethnography
European Climate Exchange, 260
evangelical Christians, 135, 249, 254, 259; on blessing the facts, 126–27, 137–38, 146–47, 154–55, 160; civic activism of, 131, 135, 140, 157–58; conversion on climate change of, 131, 146–47, 173n7, 254; Cornwall Declaration of, 132, 145–46, 257, 274n15; on evolution, 126, 147–48, 153; institutional changes among, 134–36; leader-messengers of, 9, 22, 125, 130, 136–40, 146–47,

149; moral vernacular framework of, 123–27, 140–42, 150–53, 159–61, 247, 258; political alignment of, 126, 131, 136, 153–59, 274n10, 274n12, 274n16, 275n19; Sandy Cove Conference of, 131–32, 255; scientist-partners of, 148–54, 159–60, 274n13, 275nn18–19; on social and economic issues, 129, 133–36, 141, 146, 154–59. *See also* Creation Care
Evangelical Climate Initiative (ECI), 6, 131–34, 137, 140–46, 157, 255, 274n15; media coverage of, 143, 145, 257, 274n13; opposition to, 145–46, 257; public relations management of, 143, 145
Evangelical Environmental Network (EEN), 131, 142, 255, 274n11, 274n15; "Evangelical Declaration on the Care of Creation" of, 144; founding of, 143–44; "What Would Jesus Drive?" campaign of, 131, 143–47, 254
Evangelicals for Social Action, 157
Everything Must Change (McLaren), 274n12
evolution debates, 126, 147–48, 153
expertise, 2, 4, 10, 14, 27–28, 30–31, 247–48. *See also* science experts; traditional knowledge
Exxon, 213; CSR shareholder activism and, 229; funding of skeptics by, 206
Exxon Valdez oil spill, 204, 208–12, 214

Falwell, Jerry, 157
Fenge, Terry, 68
Field Notes from a Catastrophe (Kolbert), 106
fieldwork, 5–10, 30–36; interviews in, 264n4; methodology of, 10–15; multisited ethnography in, 5, 7, 31–36; para-sites of, 281n1; study of processes and intersections in, 32–36; time context of, 15–20
financial markets. *See* Ceres; risk

Ho, Karen, 207–8, 236, 239

Hoggan, James, 103, 271n11, 280n17

Holdren, John, 113, 175, 181–82, 184, 189, 199, 205

Holland, Greg, 97–99, 102

Hopson, Eben, 267n2

Houghton, John, 131–32, 141, 146–49, 254, 255

Hubbert, Marion King, 277n10

Hulme, Michael, 4, 112–13, 166, 172–73, 199, 277n9

human rights: Inuit petition of 2005 on, 6, 8, 55, 65–70, 74–78, 254, 256, 257; politics of connection with climate change of, 70–74

Hunter, Joel, 122, 126, 127–31, 135, 149, 156–59, 273n3

Huntington, Henry, 61–62

Hurricane Katrina, 16, 95–99, 184–88, 232–36, 255

Hurricane Rita, 16, 232, 256

Hurricane Sandy, 260

ICC. *See* Inuit Circumpolar Council

ICC Canada, 48–49, 58

ICC Greenland, 48–49, 267n1

Ignatieff, Michael, 70

An Inconvenient Truth, 147, 256; on carbon dioxide, 186; discussion and critiques of, 113–15, 177; on hurricanes, 96, 185, 234–35, 280n18; Oscar of, 16; on polar bears, 99; political responses to, 103, 274n16

Indigenous People's Summit on Climate Change, 269n18

Inhofe, James, 113, 173–76, 272n19

insurance industry, 170, 276n6; coastal development and, 187–88, 236–37; costs of weather-related events to, 233–36; management of climate risk by, 236–39, 280nn20–21; vernacular of risk of, 207

intelligent design, 148

Inter-American Commission on Human Rights, 8, 65–67

interdependent worlds, 31

interdisciplinary investigation, 30–36

Interfaith Center on Corporate Responsibility (ICCR), 209, 279n5

Interfaith Stewardship Alliance, 132, 257

Intergovernmental Panel on Climate Change (IPCC), 2, 79, 144, 167, 199, 250; 2007 media initiative of, 16; adjudication of expertise by, 89–90; assessment reports of, 89, 91–92, 94, 99, 131, 141–42, 163, 179, 193, 224, 234, 257–58, 261, 275n1; critiques of advocacy by, 177–81; Emanuel's essay on, 186–87; evangelical Christians on, 148–49; formation of, 253; Nobel Peace Prize of, 16, 258; reliability of, 89; Working Group II of, 179–80

International Council for Science (ICSU), 258

International Polar Year (IPY), 43, 57–64, 258, 268n11

interpretive journalism, 108

Inuit, 249; human rights petition of 2005 of, 6, 8, 55, 65–78, 254, 256, 257; impact of resource exploration and development on, 74–78, 267n8; POPs present in bodies of, 67–68, 70, 72, 269n20; sea-level rise and relocation of villages among, 44–45, 46, 54, 160; traditional knowledge of climate change among, 46–47, 52–57, 72, 247, 268n9; vernacular for climate change among, 24, 42, 45–46, 49–52, 267n3; Watt-Cloutier's presentation of, 72–74

Inuit Circumpolar Council (ICC), 7, 8, 79–80, 143; on the Arctic Council, 77; at ASSW, 58; funding of, 69; leadership of, 267n1, 269n19; origins of, 267n2; political role of, 47–49, 63, 77–78; prioritizing by, 47–50; on resource extraction, 75–76; self-determination goals of, 8, 37, 50, 63–64, 67, 71–79, 270n21; translation and mobilization challenges of, 15, 29, 36–37, 50–52; Youth Council meeting of, 39–50

New Evangelical Partnership for Common Good, 133

A New Kind of Conservative (Hunter), 156–59

The News Gap (Boczkowski and Mitchelstein), 271n14

New York pension funds, 209

Niezen, Ronald, 71–72, 73, 75, 268n17

Nisbet, Matthew, 104, 265n13, 277n12

Nobel Peace Prize, 16, 258

Noongwook, Chester, 61

Noongwook, George, 61

Nordhaus, Ted, 255

Nordhaus, William, 275n1

Northland Church, 121–22, 127, 135

Nunavut (Canada), 49, 76

Obama, Barack, 181, 205, 259, 273n3; challenge to the public by, 243–44, 260–61; national climate action plan of, 16, 243

objective self-fashioning, 24–25

Oh, Sang Hwa, 277n12

Oozeva, Conrad, 61

Oreskes, Naomi, 4, 15, 173, 281n1; on science expert advocacy, 199; on scientific consensus, 21, 88–89, 93, 174–77, 255

Overland, James, 75, 100

ozone hole, 21, 90–92

Pachauri, Rajendra, 177

parachute journalism and anthropology, 33–34

para-sites, 281n1

Parsons, John, 164–66, 168, 170

peak oil, 182, 277n10

Penikett, Tony, 76

Pensions & Investments, 211

People's Agreement of Cochabamba, 269n18

peripheral publics. *See* alternative discourses on climate change

persistent organic pollutants (POPs), 67–68, 70, 72, 269n20

Peters, Chris, 272n17

Pew Forum on Religion and Public Life surveys, 134–36

Philosophical Investigations (Wittgenstein), 11–12

Pielke, Roger, 112–14, 277n9

polar bears, 99, 234, 259

political journalism, 119–20

political responses to climate change, 247; Ceres work on, 205, 213, 226, 239; by Clinton, 175, 238; by G. W. Bush, 3, 69, 132, 141; by Inhofe, 173–77; Kyoto Protocol and, 3, 174–75, 224, 229–30, 232, 254, 255, 260, 267n7; media coverage and, 119–20; moral and ethical frameworks for, 249–50; by Obama, 16, 243–44, 260–61; party-based contestation in, 3, 126, 131, 136, 153–59, 259, 274n10, 274n12, 274n16, 275n19; science experts and, 195–99, 247, 277n9

polling data, 3, 16–17, 42, 254, 259, 261

Polanyi, Michael, 172

Pope, Carl, 113–15

Porter, Theodore, 171–72

"Preserving & Cherishing the Earth" letter, 144

The Prince (Machiavelli), 215

Prinn, Ron, 163, 165, 276n4

Pritchard, Rusty, 133–34

Project for Excellence in Journalism, 266n16

Proofiness (Seife), 276n7

The Public and Its Problems (Dewey), 27–28

public engagement with climate change, 3–5, 243–52; of evangelical Christians, 131, 135, 140, 157–58; focus on facts and objectivity in, 28–29; future forms of, 245–46; Hurricane Katrina and, 233–36; interdisciplinary investigation of, 30–36; meaning-making in, 20–25, 171–72, 178–79, 183, 246, 265n13, 278n15; networked social relations in, 32–33; Obama's message on, 243–44, 260–61; polling data on, 3, 16–17, 42, 251, 254, 259, 261; scientific literacy

and, 3, 20–21, 23, 104, 194–95, 265n14.
See also alternative discourses on climate change; vernaculars of meaning, ethics, and morality

Pugwash Conferences on Science and World Affairs, 265n12

Pulitzer, Joseph, 108

Pungowiyi, Caleb, 52–54, 61–62

Rabinow, Paul, 34–35

Rahmstorf, Stefan, 96–98, 177

Rapp, Rayna, 265n14

Reagan, Ronald, 148

RealClimate blog, 96–98, 177

recession of 2008, 234–36, 280n19

Red Letter Christians (Campolo), 155

Reed, Ralph, 156

Reilly, John, 163

relationship building with the facts, 29–30, 50–52, 102, 125, 264n9

Rensberger, Boyce, 90–94, 104–5, 194–95, 248, 270n7, 272n15, 281n1

Reporting on Climate Change: Understanding the Science (Ward), 87–88, 93–94

The Republican War on Science (Mooney), 106, 189

Resilient Coasts Initiative, 236

Revkin, Andrew, 93, 185, 195; on an informed citizenry, 95; coverage of the Arctic by, 65–67, 99–100; *Dot Earth* blog of, 112, 114, 117; on reporting on climate change, 90, 105, 110–20, 173, 192–93, 248, 251–52, 273n24, 281n1

Rheinberger, Hans Jorg, 278n15

Richter-Menge, Jaqueline, 75, 100

"The right to be cold" (Watt-Cloutier), 65–70, 256

risk, 161, 248–52; Beck's formulations on, 115–16, 118, 170, 233, 236, 239, 250–51; Ceres approach to, 22, 38, 201, 204–8, 219–20, 224–27, 239–40; coastal development and, 187–88; cost-benefit analyses in, 165, 169–72, 198; economic recession of 2008 and, 233–36,

280n19; framing of science through, 164, 168, 177, 198–99, 247, 276n7; future forms of, 245–46; insurance industry and, 170, 187–88, 207, 233–39, 276n6, 280nn20–21; long-term view in, 207–8; media articulation of, 114–19, 273n25; Schneider on management of, 82, 273n25; scientific predictions and probability and, 98–99, 162–73, 198–99, 249, 276nn3–4

Robertson, Pat, 132, 156

Robinson, Tri, 128–30

Rockefeller family, 229

Roe v. Wade decision, 157

Rooftop Mediaworks, 145

Rosen, Jay, 31

Rosenstiel, Tom, 86, 266n17

Russia, 49, 232, 267n5

Sagan, Carl, 144

Sanyal, R. N., 210, 220

Sapir-Whorf thesis, 13, 264n9

Saving God's Green Earth (Robinson), 128

Schellenberger, Michael, 255

Scheufele, Dietram, 110, 265n13, 270n4, 272n18

Schmidt, Gavin, 96–98

Schneider, Stephen, 82–86, 91, 93, 270n1; on adjudication of expertise, 89–90; on articulation of risk, 118–19, 170, 273n25, 276nn3–4; on democratic negotiation of complexity, 251–52, 281n1; on fairness, 249; in IPCC, 89; on science and the media, 192–93, 195–99

Schrag, Dan, 173–76, 194–95, 199, 237

Schudson, Michael, 26–27, 83, 107–8, 116, 251, 266n15

Schweitzer, Peter, 267n3

Science and Technology Studies (STS), 2, 30, 178; on adjudication of expertise, 28; policy advocacy and, 181–82; on production of facts, 29; on the social in production of scientific knowledge, 197, 278n15 science experts, 7, 9–10, 28–30, 37–38, 162–200, 250; adjudication of status as, 28, 37, 83, 89–90, 105, 125, 248; advisory science of, 168–69, 276n5; blogging by, 96–97; consensus among, 10, 28, 46, 94, 107, 173–77, 184, 198; credibility challenges of, 84–87, 168; critiques of IPCC and, 177–80; media relationships of, 188–98, 277–78nn12–14; near-advocacy work of, 37–38, 112–13, 177–200, 247–48, 270n3, 277nn8–9; negotiation of moral and ethical meaning-making by, 23–25, 164; as partners of evangelical Christians, 148–54, 159–60, 274n13, 275nn18–19; policy advocacy by, 195–99, 247, 277n9; predictions and probability of risk by, 95, 98–99, 162–73, 198–99, 249, 276nn3–4; presumed impartiality and reliability of, 19–20, 28–29, 167, 183, 265n12; in Senate hearings, 173–77; skepticism and debate among, 28, 96–99, 163–65, 173–74, 277n13; translation of science for the public by, 10, 22–23, 164–65, 170, 183. *See also* traditional knowledge

scientific consensus, 10, 28, 46, 94, 107, 173, 184; IPCC's statements on, 16, 92–93; Oreskes's research on, 21, 88–89, 93, 174–77, 255; role of skepticism and debate in, 163–65

scientific literacy, 3, 20–21, 23, 104, 194–95, 265n14

scientific method, 2; coproduction of scientific knowledge and, 161, 172, 197, 264n3, 278n15; as knowledge system, 54, 56–57, 164; norms and practices of, 84–87, 172; predictions and probability of risk and, 95, 98–99, 162–73, 198–99, 249, 276n3; as process, 101, 105–6, 119–20, 274n8; *vs.* traditional knowledge, 56–64, 247, 268n9

Scopes Trial, 126, 148

Securities and Exchange Commission (SEC), 226, 240–41, 249–50

Seife, Charles, 276n7

Seiple, Bob, 131

Senate Committee on Environment and Public Works, 173–77

Shackley, Simon, 168–70, 197, 273n24

shareholder resolutions, 211–13, 217–19, 229, 239

Shishmaref, Alaska, 44–45, 46, 160

Sider, Ron, 131, 157

Sierra Club, 275n19

Silicon Valley Toxics Coalition (SVTC), 217

Singer, Jane, 32, 83

Singer, S. Fred, 92

skepticism and doubt, 15; climategate and, 17–18; of evangelical Christians, 105, 123, 131, 132, 136, 160; in industry-funded communication, 17, 92–93; of the media, 18–20, 93–95, 102–6, 248, 261, 266n17, 271n12; of the public, 3–4, 263n2; of science experts, 28, 96–99, 163–65, 173–74, 277n13; in Senate hearings on climate change, 173–77; sound science and, 132–33; teaching controversy strategy and, 89

Small Footprint, Big Handprint (Robinson), 128

Smith, Ted, 217, 219

Snyder, Alison, 223, 279n13

Socci, Anthony, 93–94, 98

socially responsible investors (SRIs), 202, 205; accountability mechanisms of, 214–15; *Exxon Valdez* and, 208–9; greenwashing and, 220–23

social media, 26–27, 143, 266nn15–16; online activism through, 16; recontextualization of information through, 30, 117–20, 188, 245–46, 251; scientist and journalist experts' use of, 96–97, 192–93, 199

Society for Environmental Journalists (SEJ), 102–3, 119–20, 271n9, 272n20

Society of Professional Journalists Code of Ethics, 83–85, 108

"The Soul of Environmentalism," 255

Speth, James Gustave, 149, 196

The State of the Arctic (Richter-Menge and Overland), 75, 100

"State of the News Media" report, 266n16

Stefansson, Vilhjalmur, 58

Stern, Nicholas, 162–68, 257

The Stern Review, 162–68, 171–72, 178, 182, 230, 236, 257, 275n1

Stockholm Convention, 68

Storm World (Mooney), 106, 187, 188

Stotts, James, 267n1

strategic essentialism, 71

strong objectivity, 34–35

Stuckenberger, Nicole, 58–60

Sullivan Principles, 209, 212, 220

Sun Company of Philadelphia, 210–11

Suncor, 220, 279n10

Swiss Re, 236, 280n20

Thin Ice exhibition, 58–60, 66–67, 74, 75, 268n12

thinkMTV, 203

Thorburn, Karin S., 230–31

350.org, 16

Time's Year of the Planet, 144

Toulou, Christopher, 236

Townsend, Solitaire, 203

traditional knowledge (TK), 46–47, 52–57, 72, 247; Arctic Climate Impact Assessment and, 54–55; collaborative approaches to, 60–64; as described in exhibitions and conferences, 58–64; development of expertise in, 61–62; ground truthing in, 54, 247; moral relationships in, 268n9; *vs.* scientific knowledge, 56–64, 268n9

translation, 13–15, 244; relationship building with the facts and, 29–30, 50–52, 102, 125; Sapir-Whorf thesis on worldview and, 13, 264n9. *See also* vernaculars of meaning, ethics, and morality

Twitter, 110, 118, 246, 272n15, 272n23. *See also* social media

Tyndall, John, 174

uncertainty, 17, 20, 73, 82, 91; in framing of climate change, 138; in reporting on climate change, 111, 116; risk and, 164, 168–69, 185, 198, 207

UN Declaration of the Rights and Duties of Man, 65

UN Environment Programme, 223, 253

UN Framework Convention on Climate Change (UNFCCC), 16, 55, 224, 250; Conference of the Parties (COP) 10 of, 254; Conference of the Parties (COP) 11 of, 65, 256; Conference of the Parties (COP) 13 of, 260; Conference of the Parties (COP) 15 of, 260; Conference of the Parties (COP) 20 of, 261; Kyoto Protocol of, 3, 174–75, 224, 229–30, 232, 254, 255, 260, 267n7

UN Stockholm Convention, 68

Union of Concerned Scientists (UCS), 183, 265n12

"An Urgent Call to Action: Scientists and Evangelicals Unite to Protect Creation," 149–50, 258

U.S. Climate Action Partnership (USCAP), 238, 281n22

U.S. Environmental Protection Agency (EPA), 258

Valdez Principles, 209–12, 214, 220, 279nn3–4. *See also* Ceres

vernacular (as term), 13, 25, 264n9

vernaculars of meaning, ethics, and morality, 1–5, 10, 23–25, 36–38, 244; in Ceres approach to climate risk, 22, 38, 201, 204–8, 219–20, 239–40, 247; coproduction of scientific knowledge in, 161, 172, 197, 264n3, 278n15; epistemological differences in, 29–30, 52; of evangelical Christians, 123–27, 140–42, 150–53, 159–61, 247; expertise and, 30, 37, 101; in expert translations of science, 10, 22–23, 164–65, 170, 183; of the Inuit, 45–46, 49–52, 79–80, 247, 267n3; journalists and, 84–87; methodology of analysis of, 10–15;

vernaculars of meaning, ethics, and morality, (*continued*)
multivocality of movements and, 134, 139; presumed impartiality and credibility in, 18–20; public understanding and meaning-making in, 20–25, 243–52, 265n13; risk and, 22, 38, 98–99, 114–19, 162–73, 248–52, 273n25; in scientific language, 54–55; translation and relationship building with the facts in, 13–15, 29–30, 50–52, 102, 125, 264n9. *See also* communication on climate change

Vineyard church movement, 128, 273n5

Virginia, Ross, 63

Wall, Kevin, 259

Ward, Bud, 87–88, 91, 93–94, 98, 117, 192–93, 272nn20–21

Warner, John, 142

Watching Weather and Ice Our Way (Krupnik et al.), 60–61

Watt-Cloutier, Sheila, 253, 258; on alarmism, 250; on climate change and human rights, 72–74, 269n20; human rights petition of, 8, 52, 55, 65–70, 74–78, 254, 256, 257; on ICC's priorities, 49–50; on resource exploration and development, 75–76; on shifting public opinion, 142–43

Way, Mark, 236

Weart, Spencer, 92–93

Weather Channel, 194

Weber, Max, 107–8

Webster, Peter, 97–99, 102

Weingart, Peter, 177–78

Wenski, Bishop, 157

The Whale and the Supercomputer (Wohlforth), 54, 268n8

"What We Know about Global Warming" (Emanuel), 186–87

"What Would Jesus Drive?" (WWJD) campaign, 131, 143–47, 254

Who Speaks for the Climate? (M. Boykoff), 98

Why We Disagree about Climate Change (Hulme), 166, 172, 199, 265n13, 277n9

Wilkinson, Katharine, 4, 133–34, 273n4

Wilson, Edward O., 149–51

Wittgenstein, Ludwig, 36; on certainty, 166; on establishment of meaning, 11–12; on language objects as forms of life, 13, 46, 244–45, 267n4; on learning language, 50, 51

Wohlforth, Charles, 54, 268n14

World Evangelical Fellowship's Unit on Ethics and Society, 144

World Meteorological Organization (WMO), 253, 258

World People's Conference on Climate Change and the Rights of Mother Earth, 260, 269n18

Wunsch, Carl, 112–13, 119, 193

Wynne, Brian, 20–21, 28, 168–70, 197, 273n24

Printed and bound by CPI Group (UK) Ltd, Croydon, CR0 4YY

27/10/2024

14580226-0004